Practising Wood in Architecture

In the stark light of the climate emergency, using wood instead of concrete, steel or masonry is increasingly seen as a way of reducing the environmental impact of architecture and construction. More and more new buildings are showcasing innovative ways to work with wood. Wood can help architects achieve ambitious sustainability targets, including the United Nations' Sustainable Development Goals.

How can architects, student architects, and those in the construction industry better understand the qualities, characteristics, and possibilities of building with wood? *Practising Wood in Architecture* explores the methods, philosophies, and possibilities of contemporary teaching practices in architecture. This book explores how architecture students are learning to build with wood and interrogates the consequences for architectural practice.

Based on original research conducted over two years, the book explores innovative projects that use wood in China, England, Finland, Germany, Mongolia, South Africa, and Switzerland. These case studies demonstrate the many advantages of wood, including its simplicity of use, its affordability, and its sustainability. The book focuses on ongoing initiatives that show the educational and professional impact of the use of wood in architecture and construction by students and professionals alike.

James Benedict Brown is an Associate Professor of Architecture at Umeå University, Sweden. James graduated from the University of Sheffield's School of Architecture in 2008 and in 2012 was awarded his PhD from Queen's University Belfast. His research interests are in critical pedagogy, architecture and wood, and design-build and live projects in architectural education, on which he has published extensively. He is the author of *Mediated Space: The Architecture of News, Entertainment and Advertising* (2018) and he is the co-editor of *A Gendered Profession: The*

Question of Representation in Space Making (2016) with Harriet Harriss, Ruth Morrow, and James Soane. Forthcoming co-authored and co-edited books include *Studio Properties: A Field Guide to Design Education* (2025) and *Architectural Thinking in a Climate Emergency* (2025).

Francesco Camilli is an architect and researcher. Between 2023 and 2025 he was Postdoctoral Research Fellow at the Norwegian University of Science and Technology (NTNU) in Trondheim, Norway, where he researched how architecture and urban design practices can generate social engagement in the transition of cities towards climate neutrality. Between 2020 and 2022 he was a Research Fellow at the Umeå School of Architecture, Umeå University, Sweden. Francesco graduated in 2016 from the Faculty of Architecture of Sapienza University in Rome, Italy, where he also obtained his PhD in 2020 from the Department of Architecture and Design. His thesis investigated participatory practices in contemporary architectural design. He has been involved in several research projects and initiatives, ranging from theoretical approaches to the affective dimension of spaces to large-scale Horizon projects linked to the New European Bauhaus initiative, and has presented and published his work internationally.

Practising Wood in Architecture

Connecting Design, Construction and Sustainability

James Benedict Brown and Francesco Camilli

Routledge
Taylor & Francis Group

NEW YORK AND LONDON

Cover image: Lapponia, Umeå School of Architecture, 2023.
© Jonas Eltes.

First published 2025
by Routledge
4 Park Square, Milton Park, Abingdon, Oxon OX14 4RN

and by Routledge
605 Third Avenue, New York, NY 10158

Routledge is an imprint of the Taylor & Francis Group, an informa business

British Library Cataloguing-in-Publication Data
A catalogue record for this book is available from the British Library

Library of Congress Cataloging-in-Publication Data
Names: Brown, James Benedict, author. | Camilli, Francesco, author.
Title: Practising wood in architecture : connecting design, construction and sustainability / James Benedict Brown and Francesco Camilli.
Description: Abingdon, Oxon ; New York, NY : Routledge, 2025. | Includes bibliographical references and index.
Identifiers: LCCN 2024037254 (print) | LCCN 2024037255 (ebook) | ISBN 9781032550817 (hardback) | ISBN 9781032550794 (paperback) | ISBN 9781003428930 (ebook)
Subjects: LCSH: Building, Wooden.
Classification: LCC NA4110 .B75 2025 (print) | LCC NA4110 (ebook) | DDC 721/.0448--dc23/eng/20240923
LC record available at https://lccn.loc.gov/2024037254
LC ebook record available at https://lccn.loc.gov/2024037255

ISBN: 978-1-032-55081-7 (hbk)
ISBN: 978-1-032-55079-4 (pbk)
ISBN: 978-1-003-42893-0 (ebk)

DOI: 10.4324/9781003428930

Typeset in Calvert
by SPi Technologies Private Limited, India (Straive)

Contents

List of Figures

Case Study Figures

Acknowledgements

The research informing this book took place over two years at Umeå School of Architecture (UMA) at Umeå University, made possible by the generous financial support of the Kempe Foundations.

After an international literature review, we approached and conducted interviews with a total of ten educators in seven countries. Those respondents gave their time and insight generously and further supported our work with prompt and diligent proofing of our transcripts. We offer our sincere thanks to Piers Taylor (Studio in the Woods), Emmanuel Vercruysse (Architectural Association), Luis Orozco and Anna Krtschil (Institute for Computational Design/Institute of Building Structures and Structural Design), Yves Weinand (IBOIS, the Laboratory for Timber Constructions at the École Polytechnique Fédérale de Lausanne), Laura Zubillaga (Aalto Wood Program), Peter Russell (University of Nottingham's Design+Build Studio), Hannes Mayer (Gramazio Kohler Research), Peter Hasdell (Insitu Project), and Joshua Bolchover (University of Hong Kong and Rural Urban Framework). We thank them, their students, and their colleagues for permitting us to reproduce the images and drawings of their projects in this book.

At Umeå University we thank Mikael Elofsson, Dean of the Faculty of Science and Technology, who first encouraged James to seek the support of the Kempe Foundations. At UMA, we are grateful for the support of two heads of department, Mikael Henningsson (2019–2023) and Cornelia Redeker (from 2023), and their deputies Michael Gruber (2019–2023) and Sara Thor and Katrin Holmqvist-Sten (from 2023). Richard Conway was a patient and critical friend in the piloting phase of our research, while Daniel Movilla Vega and Maria Luna Nobile have worked tirelessly to promote the research culture of our department. Finally, nothing at UMA would happen without Maria Bylund. Thank you all.

STATEMENT OF CO-AUTHORSHIP

We, the authors, began our collaboration as postdoctoral student (Camilli) and supervisor (Brown). We employed a methodology informed by grounded theory (see Chapter 2) to code interviews that were conducted exclusively by Camilli. As the manuscript developed, we each took responsibility for leading certain chapters, before exchanging them for review and revision. We have not ascribed individual chapters to either of us, because the final manuscript is a fully co-authored text.

Foreword

WOOD AS A TEACHING TOOL AND NOT JUST A BUILDING MATERIAL

In 2011 I travelled to Queen's University Belfast to attend the Live Projects Colloquium organised by James Benedict Brown and Ruth Morrow, not very sure of what to expect. I was delighted to find that for the first time I was in the same room as other educators who shared the same excitement about the transformational power of hands-on live projects. At the time, the field of live project/design-build/community design/service learning was fragmented and under-explored. Together we discussed the potential of this way of learning for design, recognising what we had each discovered the hard way for ourselves and, also, how much more progress we could make if we worked together.

Inspired and encouraged by this encounter, in 2012 I set up the online Live Projects Network (LPN) with my teaching partner, Colin Priest.[1] LPN tells live project stories, catalogues their attributes, and shows them in their individual contexts. We published as many case studies of live projects around the world as we could find.

We began publishing live projects online at a time when the architectural press preferred to show only high-budget and high-profile buildings. A quick look at the images of each project on the front page of LPN shows a preponderance of timber in all of its states – from raw natural forms to engineered timber products, from simple solutions to sophisticated inventions. LPN also included far more projects located in peripheral and under-served contexts than were featured in the mainstream architectural press in 2012. This was not a curatorial decision made by LPN, it is a reflection of the reality of the material and contextual strategies that hands-on projects have developed in response to the environmental, economic, and social realities that they work within.

Design methods that are common in hands-on architectural education projects are finding a place in conventional architectural practice, helped by publications such as *Spatial Agency: Other Ways of Doing Architecture* (Awan, Schneider, and Till, 2011) and also by graduates of hands-on education who are now in practice. With the Pritzker Architecture Prize being awarded to Diébédo Francis Kéré in 2022 and Yasmeen Lari winning the Royal Institute of British Architects' Gold Medal in 2023, humility and thriftiness are being recognised as effective tactics to circumvent the powerful forces of climate crisis and inequality.

Perfect timing then for this book. Hands-on architectural education has developed a broad range of design strategies that are suitable for a diverse range of contexts. Architects are more generous in their understanding of the places where design is needed and who should be included in addressing these design problems. But how can a subject as specific as 'wood in hands-on architectural education' address the complex and inter-connected issues presented by sustainable design?

This excellent book offers a unique take on the 'from farm to plate' approach to food and applies it to architectural education, offering us an alternative 'from forest to building site' pedagogy. It succeeds in the tricky task of identifying the particular qualities of wood and the forest that lend themselves so well to the challenges of learning to design sustainably.

The forest is a familiar place, an ecosystem that connects to human systems of craft, construction, design, and shelter. The forest feels simultaneously natural and architectural, both spatially and materially. We can spend time in a forest and begin to understand how natural and human systems connect and affect each other. The different species of wood that we extract from the forest are products of the climate that they grew in. We can recognise the different layers and components that a tree has formed in order to collect nutrients, reach up to the sky, and brace itself against the wind. We can use these different components to perform particular roles within buildings. Wood is light and malleable enough to allow us to work with relatively simple tools to cut, shape, and lift it, even if we are not particularly skilled.

With their readable narrative and exemplary case studies, Brown and Camilli take us on a fascinating journey, identifying wood as possessing a particularly transparent ecology. They show us how and why hands-on wood projects can be mobilised so effectively in order to teach students to design within a sustainable construction system that they can understand in its entirety. This is of huge significance to students coming of age in a rapidly changing and complex world that presents them with challenges that can easily intimidate or induce apathy. Hands-on wood projects are shown here to give novice designers access to such a specific understanding and lived experience of sustainable design that they can be empowered to contribute positive change to the systems that shape our current construction practices.

Jane Anderson
Oxford Brookes University
March 2024

Note

1   See www.liveprojectsnetwork.org.

Selected Terminology

The following definitions are provided by the authors as a compliment to the reader's dictionary to clarify or emphasise our perspectives and understandings.

Wood
1. Noun. The hard fibrous material that forms the main substance of the trunk or branches of a tree or shrub, used for fuel or timber.
2. Noun, also *woods* (plural). An area of land, smaller than a forest covered with trees.

Timber
1. Noun. Wood that has been prepared or manufactured for use in building and carpentry.
2. Count noun, usually *timbers*. A wooden beam or board used in building a house or ship.

Lumber
1. Noun (American English). Wood that has been sawn into rough planks or otherwise partially manufactured for end use.

Ecology
1. Noun. A complex web of ever-changing relations.
2. *ecological thought* the thought of complexity: see Morton (2010, 208).

Sustainability
1. Noun. The capacity to preserve a balanced ecology.

Design-build
1. Noun, also design/build, designbuild, etc. In the context of this work, a project or process in which students are involved in

both the design and construction of a building as part of their studies. *Design-build* projects in architectural education are most commonly found in North America, but are growing in popularity in other countries. Not to be confused with *design-(and)-build*, a project delivery framework, used in the construction industry, in which design and construction services are contracted by a single entity known as the design–builder or design–build contractor.

Live project

1. Noun (British English). A live project in higher education includes the negotiation of a brief, timescale, budget, and product between an educational organisation and an external collaborator for their mutual benefit. The project must be structured to ensure that students gain learning that is relevant to their educational development. See Brown (2012) and Anderson and Priest (2014).

Bibliography

Anderson, Jane and Colin Priest. 2014. 'Developing an inclusive definition, typological analysis and online resource for Live Projects.' In *Architecture Live Projects: Pedagogy into Practice*, 9–17. Abingdon: Routledge.

Awan, Nishat, Tatjana Schneider, and Jeremy Till. 2011. *Spatial Agency: Other Ways of Doing Architecture*. Abingdon: Routledge.

Brown, James Benedict. 2012. *A Critique of the Live Project*. Thesis (Ph.D.), Queen's University Belfast.

Morton, Timothy. 2010. *The Ecological Thought*. Cambridge, MA: Harvard University Press.

Chapter 1

Introduction

1.1 THE MATERIAL OF THE MOMENT

In 2019 and 2021, two timber skyscrapers were completed at almost opposite ends of the Scandinavian peninsula. For a few months, Voll Arkitekter's 85-metre tall Mjøstårnet in Brumunddal, Norway was the tallest timber building in the world. Then in 2021 White Arkitekter's 80-metre-tall Sara *kulturhus* – a mixed-use hotel, library, and theatre – opened in Skellefteå, Sweden (see Figures 1.1 and 1.2). For a moment, Scandinavia became the focus of global interest regarding the ability of tall timber buildings to confound our expectations of what can be built in wood (Wainwright, 2021). But Mjøstårnet and Sara's reign at the top of the tall timber charts was brief. In July 2022, both were eclipsed by the 25-storey, 86.6-metre Ascent Tower in Milwaukee, Wisconsin. By the time of publication of this book, the ground is scheduled to be broken in Tokyo on W350, a planned 70-storey, 350-metre mass timber skyscraper designed by Nikken Sekkei and built by Sumitomo Forestry.

Wood may be one of the oldest building materials known to humanity. Its use has been traced back to 476,000 years ago (Barham et al., 2023). But wood is also at the cutting-edge of structural, technological, and material innovation. Recent books such as *Advancing Wood Architecture: A Computational Approach* (Menges, Schwinn, and Krieg, 2016) and *Touch Wood: Material, Architecture, Future* (Ferrer, Hilderbrand, and Martinez-Cañavate, 2023) have crystalised some of the cutting-edge research into innovative applications of wood in construction. Compendia of recent projects such as *Building with Wood: The New Timber Architecture* (Toromanoff, 2023) have showcased innovative applications of wood in buildings of varying scales. Wood continues to entrance the people who design and use buildings. Perhaps it is because wood is a material with the

DOI: 10.4324/9781003428930-1

Figure 1.1 The exterior of Sara kulturhus, Skellefteå, Sweden. Photo: James Benedict Brown.

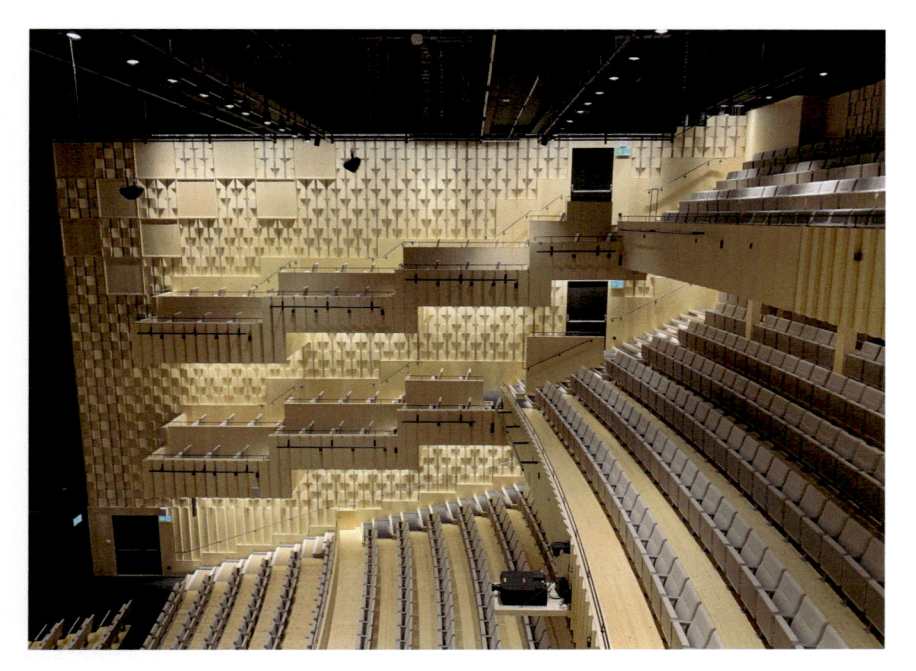

Figure 1.2 Theatre auditorium, Sara kulturhus, Skellefteå, Sweden. Photo: James Benedict Brown.

capacity to project multiple identities: rustic and nostalgic yet also able to represent modernity and cutting-edge urbanity. For advocates of biophilic architecture, wood is not only a sustainable means of construction but also a positive contributor to the mental and physical well-being of people who use buildings (Kellert, Heerwagen, and Mador, 2008).

This contemporary interest in wood architecture is set against the unavoidable backdrop of the climate emergency. The construction and operation of buildings are responsible for approximately one-third of all carbon emissions (United Nations Environment Programme and Global Alliance for Buildings and Construction, 2024). Considering the now scientifically irrefutable effects of manmade carbon emissions, wood is now considered to be one of the most promising building materials to help architects contribute increasingly ambitious sustainability targets, not to mention the United Nation's Sustainable Development Goals. The potential for wood, a renewable and carbon-absorbing material, to replace carbon-emitting concrete, masonry, and steel in construction could allow for buildings to be net carbon negative, sequestering atmospheric carbon instead of releasing more of it into the atmosphere.

It is evident that there is a resurgent interest in the potential of wood to create beautiful and sustainable buildings. How can we better understand the qualities, characteristics, and possibilities of building with wood? This book sets out to contribute to an ever-growing discourse on wood in architecture by exploring the methods, philosophies, and possibilities of teaching architecture students how to build with wood.

We have written this book with four distinct reader demographics in mind. For students of architecture, once you have begun your studies, it can be difficult to get an overview of what it might be like to study architecture in another programme or institution. National and international forums like the American Institute of Architecture Students (AIAS) in the USA or the European Architecture Student Assembly (EASA) in Europe can give you a glimpse, but very often we rely on what universities publish on their own websites to keep us informed about what is happening elsewhere in the world. International competitions like Europan, the Solar Decathlon, and the International Velux Award are good platforms for students to show off their cutting-edge work. But despite the growing interest in wood as a potentially sustainable building material, there has been no overview of some of the innovative pedagogical initiatives being undertaken in architectural education today. In addition to chapters on the qualities and characteristics of wood, this book gives you easy access to case studies of nine amazing initiatives based in (or out of) schools of architecture around the world.

For teachers of architectural design, structures, materials, and technology, this book will hopefully complement the work you are already doing by highlighting recent pedagogical innovations. If you have ever been inspired to initiate some kind of live project or design-build project in your programme, the case studies will explore some of the different models that have proven successful elsewhere. Perhaps these examples can become models or precedents that will help you win support, secure resources, and get started on your initiatives. (We look forward to talking to you before the second edition is published.)

For researchers of architecture and other design disciplines, we hope that the intellectual framework we present here provides you with inspiration for new directions in your work. We are particularly interested in reading the feedback of researchers grounded in disciplines other than our own because so many of the traditions of architectural education and practice have framed the way we work as researchers. We also hope that the case studies can be examined by other researchers after us so that the symbiosis between teaching and scholarship can be further deepened.

Finally, we have thought about the reader who practices architecture in some capacity. The education of an architect is far from complete on graduation day. The desire and the professional obligation to continuously reflect on and improve one's practice as an architect contributes to a vibrant and healthy dialogue between architectural education and practice. Perhaps you are an architect with a part-time teaching commitment. This book is also for you if you are interested in remaining connected to cutting-edge pedagogical practices in wood but also in being prepared for the next generation of architects who will work alongside you.

Whether or not you identify with one or more of these groups, a common interest that may have brought you to this book is the desire to work hands-on with wood. For students, teachers, and architects, that is often not the norm. If you are studying or teaching architecture in a higher education institution, the opportunity to build at a 1:1 scale is rare. This book gives you examples of the many different ways that an architectural education can be enriched by working hands-on with one of our most accessible, affordable, and versatile building materials.

We have tried as hard as possible, within the means of our resources and the constraints of the COVID-19 pandemic, to provide as international an outlook as possible. Some of our case studies have worked transnationally, i.e., across national borders. But a simple climatic and environmental fact has weighted our book's focus towards the northern hemisphere: most of the globe's boreal and temperate forests are found here. While we regret not being able to include case studies from other parts of the world, we are hopeful that this book will prompt a keen and critical audience to follow in our footsteps with more publications (and perhaps that second edition we are already thinking about).

There are inevitably some omissions. With limited time, resources, and pages in this book, we had to make difficult choices about which initiatives to study. Some impressive and longstanding design-build studios have employed wood for many years in North American universities – such as the Yale Building Project or Auburn University's Rural Studio – but we have found them to be well represented in the existing literature. On our own doorstep, there have been some impressive initiatives led by teachers in the design studios of Umeå University's Umeå School of Architecture (UMA), and the independent Arknat[1] programme of summer projects has delivered remarkable structures in wilderness areas in Sweden. We regret excluding these and many other great precedents from our book, but trust that our readers will find the ones we have selected a useful cross-section of best practices.

1.2 THE CONTRADICTIONS OF SWEDISH FORESTRY

Our research project was funded by the Kempe Foundations (*Kempestiftelserna*), established in memory of the Swedish industrialist Johan Carl Kempe (1799–1872) and his son Seth Michael Kempe (1857–1946). The Kempe Foundations support education and research only at institutions in Sweden's three northernmost counties of Norrbotten, Västerbotten, and Västernorrland because it was here that Johan Carl and Seth Michael made their careers speculating on forest land and operating sawmills along the Bothnian coast.

Forests and wood continue to be vital to Sweden's economy, society, and culture. Approximately 70 per cent of Sweden's land area is forested. Conducting this research in Umeå put us near some of Europe's most economically productive forests and Sweden's most successful wood enterprises. Sweden is currently the world's second-biggest exporter of pulp, paper, and sawn wood products, harvesting some 90 million cubic metres every year. Sweden produced 17 per cent of the EU's sawn wood in 2021, exporting it to customers globally. Northern Sweden is at the forefront of Europe's forest industries, but it is by no means the only country where wood is vital to the local economy (see Figure 1.3). Across all of Europe, the gross value added (GVA) of wood products – including construction timber and building materials – was €37 billion in 2020 (Eurostat, 2022), employing an

Figure 1.3 "The houses and buildings of the future are growing here" – a roadside sign near Burträsk, Sweden. Photo: James Benedict Brown.

estimated one million people (CEI-Bois, 2021). Across North America, Russia, and Asia, forestry is an equally important provider of jobs and livelihoods.

We have been able to conduct this research from a place of extreme privilege. Living in an economically successful and politically stable social democracy, we have been supported not only by a developed industrial economy but also by a research grant funded by the profits of extractive forestry.

Reflecting on the landscape which connects us with the Kempe family reveals some of the contradictions facing forestry and wood construction. The north of Sweden that Johan Carl Kempe found was very different from the one we see today. Complex old-growth forests of mixed species stretched as far as the eye could see. When the first loggers set out into the landscape, they did not replant, because they assumed the supply of wood would effectively be infinite. Sweden's industrial revolution was fuelled by an extractive approach to natural resources. In this, we discover a fascinating relationship with the so-called natural. Andreas Malm writes:

> Capital abhors the vacuum of wild nature. The capitalist class, we might recall, was brought up on hatred towards it. The pre-eminent philosopher of the plantation, John Locke, expressed the feeling eloquently: in his scheme of things, the original condition of the world was that of an unredeemed wild Common of Nature. The mission of human beings, or more precisely property-owning human beings, was to abolish that condition. The wild common ought to be enclosed, rendered productive, improved – in short, converted into a fount of profit.
>
> (2020, 76–77)

Today, our attitudes towards forests are changing. The ecosystems that forests support are becoming recognised as being important to our understanding of the climate emergency. In 2009, the 15th Conference of Parties to the UN Convention on Biological Diversity (COP15) demanded global attention be paid to the danger of ignoring the loss of biodiversity associated with damage to natural ecosystems. In 2024, the member states of the European Union passed legislation to restore at least 30 per cent of forests, grasslands, wetlands, rivers, lakes, and coral beds by 2030, increasing to 60 per cent by 2040 and 90 per cent by 2050. The architecture and construction industry cannot ignore these concerns. We can reduce our carbon emissions by using less energy- and emissions-intensive materials like steel and concrete, and more wood that can sequester atmospheric carbon for the lifetime of a building. We can also reduce the volume of petrochemicals used in wood construction and the manufacture of wood products, developing alternatives to chemical adhesives and designing buildings not only for sustainable assembly but also for sustainable disassembly. But we also have an obligation to understand the forests which produce the products feeding our wood revolution.

Tension now exists between advocates for industrial forestry and their critics. The growing demand for wood threatens Europe's forests and the biodiversity they support. Forests are regarded as 'the lungs of Earth' (Macron, 2019) and their

Figure 1.4 A nineteenth-century Swedish rundloge in Västerbotten county. These log barns were used for threshing and agricultural storage. Photo: James Benedict Brown.

destruction as 'a crime against humanity' (Goldberg, 2019). These criticisms are easy for European politicians and campaigners to make when the subject is South American rainforests, for example. But here, closer to home, one-quarter of Sweden's unprotected old-growth forest has already been lost to commercial forestry in only 16 years. At the current rate, what remains will be lost in the next 50 (Ahlström, Canadell, and Metcalfe, 2022).

The forest industry has long sought to reassure customers that its products are sustainably farmed with internationally recognised standards – for example, those administered by the Forest Stewardship Council (FSC) or the Programme for the Endorsement of Forest Certification (PEFC). But these hallmarks have been criticised by organisations such as Greenpeace, the Wilderness Society, and Stichting Fern because of the environmentally damaging practices they continue to permit, not least the destruction of old-growth forests and the creation of man-made monocultural woodland plantations. Here in Sweden, criticism of the industrial practices of clear-cutting forests and replacing them with more profitable monocultural plantations has become increasingly widespread. The relationship of the academy to industry has also been criticised. In a series of articles for the daily newspaper *Dagens Nyheter* and the book *Skogslandet* ('*The forest country*'), investigative journalist Lisa Röstlund has shown how supposedly independent academic researchers and the scientific agenda of the *Future Forests* project (2009–16), hosted by the Swedish Agricultural University and funded by the Swedish Foundation for Strategic Environmental Research, were influenced by stakeholders in the forest industry (Röstlund, 2021, 2022); The emotive and cultural ties to the forest felt by people who live and work in it were further explored by Maciej Zaremba, who prompted widespread public debate about forest governance, clear-cutting, and monocultural replanting in his collection of essays *Skogen vi ärvde* ('*The forest we inherited*', Zaremba, 2012).

These conversations are not easy to have. The forest is owned by some 320,000 different forest owners, including individuals, limited liability companies, pension and investment funds, and the Swedish state. Exercising *allemansrätten* (the right to roam freely granted in Nordic countries), we can walk through a contiguous forest and step without any awareness from the land held by a multinational investment fund to a plot owned by an individual planning for their retirement. The economy of Sweden is dependent on forestry both directly – through employment and economic activity – and indirectly – through the maintenance of economically productive forests by hunting associations and through the investment of personal and collective pension funds in forest plantations. Any critique of how wood is harvested, manufactured, and used brings our research into conflict with normative attitudes about the sustainability of wood and forest industries.

1.3 OUR SWEDISH OUTLOOK

This book emerges from a two-year research project at UMA. UMA is Sweden's northernmost school of architecture, located more than 600 kilometres north of Stockholm at 63.5° north. We are at the same latitude as Reykjavík in Iceland, Yellowknife in the Northwest Territories of Canada, and Murmansk in Russia. Our winters are long, cold, and dark, with just three and a half hours of daylight on the winter solstice. Several metres of snow normally accumulate in the winter.

At UMA, we turn ourselves inwards during the autumn term. We work in a building designed for us by the Danish architect Henning Larsen, situated beside the Ume River on the university's Arts Campus. Around lunchtime on the darkest winter days, we enjoy the many hundreds of windows that bring shafts of near-horizontal light into our studios for just a few minutes (and some of our bravest students go ice bathing in the river, see Figure 1.5). When we go home, we cycle and walk over thick ice and in temperatures of –20° or lower. Our summers are short but filled with daylight. By the time our students finish their final presentations, they celebrate with a swim that no longer requires a hole in the ice. In the final weeks of the semester, our first-year students build small pavilions in the park outside (see Figure 1.6). Then, when the traditional Swedish summer vacation begins around midsummer, many of us retreat to the lakes and rivers so that our skin can absorb the midnight sun and feed the mosquitoes.

Figure 1.5 UMA students bathing in the Ume River, Umeå, November 2023. Photo: James Benedict Brown.

Figure 1.6 UMA first-year students constructing a pavilion at the end of the spring semester, Umeå, Sweden. Photo: Navid Ghafouri.

Although we are proud to be an international school, teaching and operating in English with a multinational staff, we are rooted in a landscape and an environment unlike that of any other school of architecture. While life in the north of Europe is idyllic, the effects of manmade climate change on the sub-Arctic are deeply worrying. Longer, hotter, and dryer summers are causing more frequent forest fires. The Gulf of Bothnia, which separates Sweden from Finland to the east, no longer freezes over. In the 1960s, it was possible to drive a car across the sea from Västerbotten in Sweden to the Swedish-speaking Ostrobothnia (*Österbotten* or *Pohjanmaa*) in Finland. Researchers at the Swedish Agricultural University (SLU) research station at Svartberget in Vindeln have found that northern Sweden has effectively lost a month of winter in the last forty years (SVT Nyheter Västerbotten, 2023). Meanwhile, to the north of us, researchers have found that the Arctic has warmed nearly four times faster than the rest of the planet (Rantanen et al., 2022).

As a university, we are not blameless. In the years leading up to the COVID-19 pandemic, nearly the entire volume (slightly more than 99.5 per cent) of carbon dioxide generated by Umeå University's travel came from commercial flights. Choosing between a six-and-a-half-hour daytime train, a nine-hour night train, or a 50-minute flight, most of our staff, students, and visitors choose one of the three airlines that fly from Umeå to Stockholm. Ironically, Sweden's diminishing winter climate continues to be a driver of tourism, itself a contributor to climate change. And, after two cancellations in 1990 and 2016 due to a lack of snow and ice, the Swedish Rally (part of the World Rally Championship) moved permanently north from Värmland county to Västerbotten county in search of reliably colder winters suitable for motor racing in snow and ice. We now welcome more than two hundred thousand spectators and participants, many who fly here, just to watch predominantly fossil-fuelled cars race through the forests around Umeå.

Northern Sweden is a politically stable and economically prosperous region. But Amitav Ghosh warns us that 'climate change poses a powerful challenge to what is perhaps the single most important political conception of the modern era: the idea of freedom' (Ghosh, 2016, 119). Here in the north of Europe, that individual freedom is often manifested in activities with high carbon intensity: leisure and sporting activities on vehicles with internal combustion engines, flights to warmer climates in the winter, and a year-round supply of fruit and vegetables imported from as far away as the Global South. By acknowledging the need to arrest manmade climate change, we also acknowledge that individuals in places like ours need to critically interrogate every aspect of our lives. Within the limited parameters of this book, the education and practice of future architects have been our primary concern.

1.4 INTO THE WOODS

Bruno Maçães writes that we are at 'the moment when our fundamental way of relating to the natural environment is rethought and, as a result, new political and economic arrangements become both possible and necessary' (Maçães, 2021, 119). It is not enough for us to just build more in wood, we need to understand wood not as an inert material but as a living product taken from living ecosystems. To do this, we as architects, educators, and students must interrogate the ways in which we understand – and misunderstand – our conceptions of the sustainability of wood.

Figure 1.7 Skuleskogen National Park, Västernorrland county, Sweden. Photo: Francesco Camilli.

We set out to write a book about the hands-on use of wood in architectural education, but what we found has implications for architectural practice and research. So, while this book originates in research into architecture's many innovative pedagogies, in every case study we were reminded of the vast number of future architects who were being exposed to alternative methodologies and ideologies regarding wood. This book is rooted in architectural education, but like everything we do in schools of architecture, it looks towards future architectural practice.

In the following chapter, we explore Western historical and cultural metaphors of the forest as a liminal space of possibility, exploration, and adventure to illustrate the relationship between architecture and timber and how this is reflected in education. We contrast these attitudes with indigenous worldviews which regard the forest not as something 'other', but as part of a holistic environment. Then, to adequately articulate the sheer variety and diversity of innovations in the field today, the remainder of this book alternates between case studies and chapters that explore various aspects of wood as a material (Chapter 3), designing with wood (Chapter 4), teaching with wood (Chapter 5), the ecology of wood (Chapter 6), and the politics of wood (Chapter 7).

We identified useful case studies from almost every level of architectural education, both undergraduate and graduate. But rather than focus on just one type of pedagogical initiative, such as the design-build project, we looked for examples of how students of architecture can engage in wood in a variety of different ways, durations, and formats. Our case studies vary in duration from a long weekend to an entire year of study. One is independent of any higher education institution, attracting students of architecture during their summer vacation who can afford a fee for participation and full board. Others function as electives within design studios or larger programmes. One is an internationally renowned full-time graduate programme of 60 ECTS (European study credits) charging students in excess of €8,000 a year in tuition. These case studies, therefore, cover projects, studios, courses, and programmes. We choose to refer to them as *initiatives* to respect the diversity of our case studies' constitutions and to emphasise the very great imagination and creativity displayed by all of them. On occasion, we refer to the case studies out of their sequence in the book, so you may find yourself flipping forwards and backwards to find them.

Note

1   See https://arknat.com/.

Bibliography

Ahlström, Anders, Josep G. Canadell, and Daniel B. Metcalfe. 2022. 'Widespread unquantified conversion of old boreal forests to plantations.' *Earth's Future*, 10(11). https://doi.org/10.1029/2022EF003221

Barham, L., G.A.T. Duller, I. Candy, C. Scott, C.R. Cartwright, J.R. Peterson, C. Kabukcu, M.S. Chapot, F. Melia, V. Rots, N. George, N. Taipale, P. Gethin, and P. Nkombwe. 2023. 'Evidence for the earliest structural use of wood at least 476,000 years ago.' *Nature*, 622(7981). https://doi.org/10.1038/s41586-023-06557-9

CEI-Bois. 2021. 'About.' Accessed 10 November 2023. https://www.cei-bois.org/

Eurostat. *Wood Products – Production and Trade*. Accessed 19 December 2022. https://ec.europa.eu/eurostat/statistics-explained/index.php?title=Wood_products_-_production_and_trade

Ferrer, Carla, Thomas Hilderbrand, and Celina Martinez-Cañavate. 2023. *Touch Wood: Material, Architecture, Future*. Zurich: Lars Müller Publishers.

Ghosh, Amitav. 2016. *The Great Derangement: Climate Change and the Unthinkable*. Haryana: Penguin Books.

Goldberg, Beverly. 2019. 'Marina Silva: "the fires in the Amazon are a crime against humanity".' *openDemocracy*. https://www.opendemocracy.net/en/democraciaabierta/marina-silva-los-incendios-en-el-amazonas-son-un-crimen-de-lesa-humanidad-en/

Kellert, Stephen R., Judith Heerwagen, and Martin Mador. 2008. *Biophilic Design: The Theory, Science, and Practice of Bringing Buildings to Life*. Hoboken, NJ: John Wiley & Sons, Inc.

Maçães, Bruno. 2021. *Geopolitics for the End Time: From the Pandemic to the Climate Crisis*. London: Hurst & Company.

Macron, Emmanuel. 22 August 2019. Twitter/X. See https://bit.ly/3eCJ54F

Malm, Andreas. 2020. *Corona, Climate, Chronic Emergency: War Communism in the Twenty-First Century*. London and New York: Verso.

Menges, Achim, Tobias Schwinn, and Oliver David Krieg. 2016. *Advancing Wood Architecture: A computational Approach*. London: Routledge.

Rantanen, M., A.Y. Karpechko, A. Lipponen, K. Nordling, O. Hyvärinen, K. Ruosteenoja, T. Vihma, and A. Laaksonen. 2022. 'The Arctic has warmed nearly four times faster than the globe since 1979.' *Communications Earth & Environment*, 3(1).

Röstlund, Lisa. 2021. 'Forskare: Skogsindustrin styrde vår forskning.' *Dagens Nyheter*. https://www.dn.se/sverige/forskare-skogsindustrin-styrde-var-forskning/

Röstlund, Lisa. 2022. *Skogslandet: en granskning*. Stockholm: Forum.

SVT Nyheter Västerbotten. 2023. 'SLU:s nya rön: En månad kortare vinter vid Svartberget i Vindeln.' *SVT Nyheter*. Accessed 13 March 2023. See https://www.svt.se/nyheter/lokalt/vasterbotten/en-manad-kortare-vinter-vid-svartberget

Toromanoff, Agata. 2023. *Building with Wood: The New Timber Architecture*. Munich: Prestel Verlag.

United Nations Environment Programme and Global Alliance for Buildings and Construction. 2024. *Global Status Report for Buildings and Construction – Beyond Foundations: Mainstreaming Sustainable Solutions to Cut Emissions From the Buildings Sector*. See https://wedocs.unep.org/20.500.11822/45095

Wainwright, Oliver. 2021. 'Isn't it good, Swedish plywood: The miraculous eco-town with a 20-storey wooden skyscraper.' *The Guardian*. Accessed 28 April 2022. See https://www.theguardian.com/artanddesign/2021/oct/14/skelleftea-swedish-plywood-eco-town-20-storey-wooden-skyscraper-worlds-tallest

Zaremba, Maciej. 2012. *Skogen vi ärvde*. Stockholm: Weyler.

Chapter 2

The Architect and the Forest

2.1 SEEING THE WOOD FOR THE TREES

When the Vikings arrived in Iceland, more than a thousand years ago, forty per cent of the surface area of the island was forested. In as little as three hundred years, almost all of the trees in Iceland had been cut down for farming, fuel, and construction. Today, despite intensive reforestation efforts, less than two per cent of Iceland is forested.[1]

Forests, the idea of forests, and even (in the case of Iceland) the absence of forests are central to many Occidental cultures. The idiom *to see the wood for the trees* (sometimes *to see the forest for the trees* in American English) – used most often in a negative construction – refers to the perception of a pattern from a mass of detail. *Not* being about to see the wood for the trees implies that the subject is so immersed in an environment that they are unable to stand back and see the broader situation or the bigger picture.[2] In Figure 2.1, cartoonist Wilbur Dawbarn plays with both the idiom and the deeper meaning behind it: that we can very easily lose sight of the bigger environment when we focus on the constituent parts. From Icelandic history we can learn about the dramatic consequences of Viking settlers – they discovered an island of seemingly endless forests with no conception of the ecological damage that would be caused by deforestation much later.

As we will explore later, scientific researchers often talk about their 'field' of study. In our case, we choose to describe a 'forest' of study. To be immersed in a forest, surrounded by trees, is to be immersed in a complex environment that is too big to be

DOI: 10.4324/9781003428930-2

Figure 2.1 Not seeing the trees for the wood. Drawing: Wilbur Dawbarn and *Private Eye*.

perceived from any one vantage point. It requires navigation, exploration, and movement. To understand the bigger picture, we need to be moving continuously, triangulating ourselves, and seeing how the parts relate to the whole. We also need to tread carefully, since the forest is not an unlimited resource.

The next two parts of this chapter extend this metaphor of the forest in two ways: firstly, to elucidate the conceptual approach of the book, and secondly, to explain our research methodology.

2.2 GOING INTO THE FOREST: A CONCEPTUAL APPROACH

Brüder Grimm
Hänsel und Gretel

Otto Kubel pinx

Figure 2.2 The woodcutter and his wife leave Hansel and Gretel in the forest. (Otto Kubel, 1922 or earlier). Print: The Jack Zipes Historic Fairy Tale Postcard Collection and Minneapolis College of Art and Design.

In Western literature, the forest is often depicted as a liminal space. It is a space of fairy tales and parables, a domain of possibility, danger, and illicit behaviour: 'in the religions, mythologies, and literatures of the West, the forest appears as a place where the logic of distinction goes astray' (Harrison, 1992). Catherine Addison (2007) argues that in European and North American literature – landscapes which were heavily forested before the arrival of humankind – forests can be seen in three different but interrelated ways. Firstly, there is a primal view of the forest which contrasts with the village, town, or other built environment. Secondly, there is the place of wandering or error, in which the protagonist can become lost, exposed to danger, or potentially encounter magic. This place of trial or tribulation is defined not in opposition to the built environment but in terms of the forest's own mythology and mystery. Thirdly, a forest may be viewed as a place of safety for the outlaw or exile.

The story of *Hansel and Gretel* appears to originate in central Europe's Late Middle Ages. Transcribed and published by the Brothers Grimm in 1812, the story tells of a brother and sister who are abandoned in a forest and kidnapped by a witch. Gretel kills the witch, saving her brother, and then the two siblings steal the witch's possessions and escape back to civilisation. Likewise, in the tale of *Little Red Riding Hood*, the protagonist has been given strict instructions to stay on the path that leads safely through the forest to her grandmother's house. A sly wolf, a native inhabitant of the forest, deceives her and attempts to eat her. In various versions of the story, the female protagonist is saved by either a lumberjack or a hunter: male heroes whose profession is associated with man's conquering of the wild and untamed forest.

Stories like *Hansel and Gretel* and *Little Red Riding Hood* reveal the fearful relationship between European society and the forest. After the First Agricultural Revolution, a large part of society shifted from the nomadic state of hunting and gathering food to a more sedentary means of pastoral agriculture. This led to the domestication of plants and animals and the beginning of an agrarian lifestyle. With this came the development of larger human settlements where trade could take place, and the birth of industrial society. Out of these eras came a more oppositional view of human society and the forest: a village is a place of safety and civilisation, while the forest is a domain of danger and wilderness.

Away from the safety and certainty of human settlements, the forest is a place where it is easy to get lost. The forest obscures our view and confuses our navigation. The idiom *to see the wood for the trees* further illustrates the way in which European culture tends to regard the forest as a space apart: something bigger than us which cannot be comprehended in its entirety without first stepping away from it. As a result, the forest often confounded those who saw it as a landscape that must be tamed. In the process the forest became actively *othered*. This – primarily Western – *othering* of the forest happened in two stages: firstly, with the Enlightenment, and secondly, with the Industrial Revolution.

In 1620, Francis Bacon (1561–1626) referred to humankind's 'divine bequest' to exercise control over nature. The French philosopher René Descartes (1596–1650) wrote in his *Discourse on Method* (1637) that 'science should make us masters and possessors of nature.' Descartes' Enlightenment vision of human mastery and possession of nature reached maturity in the eighteenth and nineteenth centuries when, during the Industrial Revolution, 'forests become the object of a new science of forestry, with the State assuming the role of Descartes's thinking subject' (Harrison, 1992, 108). This *othering* of the forest continued into objectification, as forests were surveyed and calculated in terms of their volume of disposable wood.

Even when Enlightenment philosophers recognised the non-commercial value of the natural world, they tended to do so from an anthropocentric point of view. In *On Nature* (1874) John Stuart Mill (1806–73) explored the concept of nature and the role of nature in shaping human morality and ethics. For Mill, studying nature was a means to gain insights into the principles of ethics, as well as the basis for understanding the role of human beings in the world. In this sense, Mill was just one of many philosophers who could not relinquish an anthropocentric view of the natural world: something that existed to educate, inform, and serve humankind. Perhaps the first person to offer a non-anthropocentric view of the natural world was the German geographer, naturalist, and explorer Alexander von Humboldt (1769–1859). Through his travels in Latin

America and his assiduous qualitative observations and experiments, Humbolt recognized the interconnectedness of different aspects of the natural world, laying down the first argument for human comprehension of ecosystems as complex systems. Humbolt influenced a generation of ecological thinkers, including Charles Darwin, Henry David Thoreau, and Ernst Haeckel. But the servitude of nature to mankind was hard to escape. Even Henry David Thoreau (1817–1862) famously went to live a simple life beside Walden Pond because he saw the natural world first and foremost as a source of spiritual nourishment and enlightenment for the human mind. The woods, for Thoreau, were a chance 'to live deliberately, to front only the essential facts of life, and see if I could not learn what it had to teach, and not, when I came to die, discover that I had not lived' (Thoreau, 2014, 67).

With industrialisation, the forest took on a new role: a natural world that was deliberately othered from the human world. This attitude towards nature is not limited to capitalist societies. In Maoist China, nature was something to be conquered, droughts were battles over which victory had to be secured, and insects, rodents, and even birds were to be wiped out (Shapiro, 2001). The very idea that there is such a thing as raw, untouched nature is central to the ideology of industrialisation because it is foundational to the post-industrial conception of liberty. Writers such as Dipesh Chakrabarty and Amitav Ghosh have argued that philosophers of freedom were 'concerned with how humans would escape the injustice, oppression, inequality, or even uniformity foisted on them by other humans or human-made systems' (Chakrabarty, 2021, 208). In *The Great Derangement*, Ghosh explains 'nonhuman forces and systems had no place in this calculus of liberty; indeed being independent of Nature was considered one of the defining characteristics of freedom itself' (Ghosh, 2016, 119). Ghosh's argument is that climate change challenges and interrogates the idea of freedom, perhaps 'the single most important political conception of the modern era' (2016). Philip Ursprung proposes that a way for architects using wood to escape this ideology is to:

> … put ourselves in the place of wood and share its perspective. If we perceive wood not as an object of aesthetic contemplation nor as an object of consumption, but as a subject in its

own right, this can help us change our perspective. Rather than seeing wood as a renewable – and thus potentially endless resource, we should care for it as fragile, vulnerable, and precious.

(Ursprung, 2023, 18)

William Cronon identifies a paradox in the post-industrial conception of the natural world. If we accept that there is such a thing as a natural world, then our presence in it represents its downfall, one in which we cannot hope to find salvation from the impending climate emergency. He writes:

…if by definition wilderness leaves no place for human beings, save perhaps as contemplative sojourners enjoying their leisurely reverie in God's natural cathedral—then also by definition it can offer no solution to the environmental and other problems that confront us. To the extent that we celebrate wilderness as the measure with which we judge civilization, we reproduce the dualism that sets humanity and nature at opposite poles. We thereby leave ourselves little hope of discovering what an ethical, sustainable, honorable human place in nature might actually look like.

(Cronon, 1995, 81)

The first waves of industrial forestry came to northern Sweden in the eighteenth century and ravaged the landscape, as gangs of lumberjacks sawed trees with no consideration for the consequent damage to the environment and without a remediation plan. As forestry developed into a more organized industrial activity, strategies for replanting woodland emerged, including the economic preference for monocultural replanting of single-species plantations that would deliver a more uniform harvest of wood with consistent sizes and characteristics.

Since indigenous societies did not make the same shift as Western peoples from nomadism to sedentary agriculture, the indigenous worldview of the forest differs substantially from the European one. The forest is not a space apart, a liminal space, or a place of danger that children must be warned about. The forest, and the animal and plant life within it, are not *othered* – they are

Figure 2.3 Deforested land near Järvsjö, Västerbotten county, Sweden, c. 1920. Photo: Hjalmar Höglund, Collection of the Västerbotten Museum.

part of the human world, and the human world is part of them. Today, we are writing this book in one of the world's most economically important forest landscapes, responding to an increased public demand for renewable building materials, packaging, and biofuels. But it is also a site of conflict between two worldviews of the forest: Sámi and European. Working in northern Scandinavia and in southern Sápmi, we authors have the possibility to confront our Euro-centric biases, blind spots, and prejudices. But we also have an opportunity to interrogate the region's history of confrontation between extractive carbon industries and indigenous ways of living in the landscape. In this book, we regard forestry as an extractive and colonial endeavour: one which regards the forest as a resource to be mined and exported to satisfy far-removed interests, while disregarding the impacts of these actions on the local ecological context. The lesson for us, as researchers, is that if we enter the forest with the same uncritical assumptions, we risk treating it in an extractive way. For this reason, we have explored practices that promote a renovated approach to the forest and the materials it produces, establishing relations of care and respect.

2.3 NAVIGATING THE FOREST: A METHODOLOGICAL APPROACH

In the third and final section of this chapter, we provide the reader with a brief overview of the research employed in this work. As our research addresses the decisions made by teachers and students in architectural education, qualitative research methods were used to select and analyse our case studies. This section of the chapter was developed, in part, through several iterations of written explanations of our work for scientific journals, which tend to demonstrate a higher degree of methodological scrutiny than one might expect to find in a book such as this one. We include this section for the benefit of students and researchers to demonstrate the link between our conceptual and methodological approaches.

Scientific researchers often refer to a sphere of interest as their 'field' of study. Perhaps a 'forest' of study is more appropriate for our endeavour. A 'field' suggests open land over which the fieldworker can enjoy a commanding vision. As we have seen in the preceding section, western society tends to treat forests as places of obscurity and disorientation. Like any researcher, we approached our 'forest' of study with an initial understanding of what we could perceive, and we set out on our journey into the forest with certain assumptions based on what we knew or believed. We did not have a roadmap, but we did have a sense of our intended direction. For the benefit of the reader, it is important to re-state what might seem obvious: we are both white, male Europeans educated as architects but now establishing ourselves as researchers in the academy thanks to doctoral degrees. We both have some experience in architectural practice but are now both more firmly located in the academy, where progression and reward follow scientific publication, securing research funding, and delivering high-quality teaching. Finally, we come from England and Italy, speak a total of five European languages, and now both live and work as immigrants in different parts of Scandinavia. We both possess cultural capital that made it easier to establish ourselves here, in a landscape that is in no small part defined by its forests and the economic activity derived from them.

To survey the 'forest' of study, we undertook an initial literature review, using keywords related to our initial interest in initiatives that provide first- (Bachelor) or second-cycle (Master) architecture students with opportunities to engage with the realization of a wooden building or structure as part of their studies. The process of identifying and gathering data on these potential case studies was not immediately straightforward. The literature about such projects is not homogeneous: some initiatives have been the subject of books and publications, some are publicised with videos or other media online, whereas others barely have a website. Through cross-referencing of literature published in edited and/or peer reviewed publications (books, journals, magazines, etc.) with non-reviewed publications (websites, social media posts, etc.) we eventually identified eight case studies operating on three continents. A ninth case study was identified by chance later in our research and, after a critical evaluation of its characteristics, was added to the survey.

In order to better understand initiatives in architectural education that involved students working hands-on with wood, we could choose to study the literature derived from or describing these case studies, or we could choose to talk to the people involved. Having some familiarity with the technique (see Brown, 2012), we decided to set up interviews to speak to the teachers directly involved in the case studies. We could have chosen to speak to students or graduates of these initiatives, but since students tend only to experience these initiatives once (and rarely have the opportunity to compare them) we deliberately chose to speak to teachers.

In parallel with the identification of potential case studies, we drafted and piloted an interview schedule. This was made possible thanks to the participation of a teacher colleague who has experience tutoring students working hands-on with wood in a small pavilion project that takes place every summer at our institution. We revised the interview schedule around five topics: timber as a material, design and architecture, education, ecology, and politics. This interview schedule was supplied in advance to our interviewees, and started by inviting them to respond to a series of assertions developed from our initial review of the literature.

In our preceding research into architectural education (Brown and Russel, 2022), we had found that the terms 'live project' and 'design-build project' are sometimes used interchangeably. In fact, these are different kinds of initiatives, and the literature would suggest that they come from two different pedagogical traditions. Adopting the attitude of explorers entering the forest of study, we made the assertive first step of clarifying the definitions, and sharing these in advance with our interviewees.

> We define a design-build project in architectural education as a project involving students in the actual construction of a structure or building as part of their studies.
>
> We define a live project in architectural education as a project that involves students in delivering a real outcome to a real client according to a real budget.

Drawing on the same literature review, we also shared in advance with the interviewees a list of assumptions about the use of wood in architectural education.

> For architecture students, being involved in the realisation of an actual building can make more tangible the nature of architectural design and construction.
>
> Design-build projects and live projects can give evidence of this assumption, together with projects realised within research centres.
>
> Wood is a versatile material that can be used in many ways and at different levels of technological complexity, from the simplest to the most advanced, and this versatility makes it particularly suitable to be used in design-build/live projects
>
> Wood has the potential to be more sustainable than steel, concrete, or masonry.
>
> The ecology of wood is a complex system that involves many different actors: forests, fauna, human communities, economies, politics: therefore, we want to avoid reductionism when approaching this topic.
>
> We think that the use of wood in architectural education initiatives like design-build/live projects can be a strategy to

effectively promote sustainability among students in architecture and, consequently, in the actual practice of architecture.

We invited our interviewees to read and respond to these assertions before we began our interviews, allowing for clarifications and corrections. Interviews with one or two of the identifiable leaders of each case study were held to obtain some degree of qualitative information not only about their methods and outcomes but also about the actual experiences, unexpected outcomes, social interactions, and impacts of working with wood.

The interviews were advantageous because they provided access to insights on the work of the different initiatives and institutions that might not otherwise be found in the literature, especially that which is published by the institution itself for the purpose of marketing and publicity. This allowed us to understand in which way these initiatives relate to timber from an ecological point of view, and therefore what kind of technical, social, or environmental relationships they try to explore. The interviews were conducted remotely, via video link, and a recording was transcribed, proofed, and returned to the interviewees for approval. We then used a four-step process of coding to build a grounded theory from the interviews (Charmaz, 2014). Throughout this process, we treated our interviews as neither wholly objective nor wholly subjective, rather as 'intersubjective' collaborative constructions of the interviewer and interviewee (Laing, 1967).

In the first step, single significant phrases in the interviews were marked with a 'code,' i.e., a short sentence aimed at anchoring, summarising, and synthesising them. The second step consisted in assembling the codes from across all of the interviews. To give just one snapshot of the coding process, in Table 2.1 we show how some of the codes generated from two interviews (X & Y) were subsequently grouped together under five concepts relating to the category 'education.' Table 2.1 is just a snapshot: our interviews of approximately one hour each generated between 80 and 180 codes, which were subsequently grouped according to five themes.

Each of the codes was generated from a passage in the interview transcript. To illustrate one code, the code 'Building with

Table 2.1 Codes and concepts generated from interviews X and Y grouped under the theme 'education'

Concepts	Category: Education				
	Educational timber	**Hands-on learning**	**Ecological awareness**	**Alternative education**	**Multidisciplinarity**
Interview X	Using wood because it is sustainable Building with wood makes architecture more tangible Timber as a collaborative research material	Thinking about fabrication during design	Sustainability through reduction of the use of materials Using more wood is not necessarily environmentally better	Educating researchers Critique of architectural education: does not link spatial quality to buildability Teaching that thinking about manufacture during design makes better architecture	Including not only architecture and engineering students Design/build as an interdisciplinary project
Interview Y	Timber becomes a studio theme Quick training for students Timber is easy to move and lift	Addressing the fact that most students have not ever built anything in their life	Students seek social impact Students do not consider timber a natural material	Commitment makes the difference with normative architecture	Contribution of expert teachers

wood makes architecture more tangible' associated with interview X was derived from the following statement.

> So, I completely agree with the first assumption: I think that not only is being part of the realization of an actual building project important within the academic sphere, but also if students spend time in industry and then return to academia. *This makes more tangible what it is that architecture actually is to them.* I fully agree with the first assumption, which basically ties directly into the second one.

Codes were compiled separately from their interviews and then transposed to Post-it notes for a series of coding workshops. While powerful software exists to facilitate this process, like many architects we tend to work better in tactile, three-dimensional media, so we co-opted the largest wall available to us in the Umeå School of Architecture and went to work (see Figure 2.4).

Figure 2.4 Using sticky notes to visually cluster codes. Photo: James Benedict Brown and Francesco Camilli.

Practising Wood in Architecture

In the next stage of our grounded theory, codes started to form clusters around 'concepts.' As the codes described in this small example started to cluster together, five concepts emerged from this coding process: educational wood, hands-on learning, ecological awareness, alternative education, and multidisciplinary collaboration.

Firstly, the concept 'educational wood' immediately resonated with one of the main goals of our research: to have a better understanding of how the use of wood can affect the methods and the efficacy of architectural education. For this reason, in the interviews we tried to provide open questions that allowed the interviewees to articulate the reasons and the consequences of their use of timber. Codes that emerged here were those that most closely defined the educational properties of the material.

Secondly, in terms of 'hands-on learning,' design-build projects and live projects are educational initiatives whose purpose is to give students a real-life experience of the architectural process. The interviews included questions about this aspect, with the aim of obtaining examples showing the actual efficacy of this approach. All the interviewees were concerned with the lack of practical experience for students and trying to give them an awareness of the complex material consequences of spatial design choices. Codes that clustered around this concept were expressions of thoughts or ideas precisely about that hands-on tactile experience.

Thirdly, in terms of 'ecological awareness,' one of the speculative outcomes of our research is that timber has some kind of 'transparent' ecology that is easier to grasp compared to other building materials. This concept emerged through the interviews in different ways: timber is not only considered a sustainable material in itself, but also as a material that is part of an ecology involving many other actors. All the case studies promote a deep and critical understanding of wood in order to enhance its sustainable properties and to promote its better use.

Fourthly, in terms of defining paradigms of 'alternative education,' design-build projects and live projects involving a hands-on experience of construction are, by definition, initiatives that set themselves apart from normative architectural education.

This is the case also in our case studies, all of which claim to be different from traditional ways of teaching architecture. The interviewees are generally critical of normative architectural education, in particular of its lack of focus on practical aspects of design, like fabricability, or its tendency to approach the material aspect of a building in an abstract way. This concept emerged from the multitude of expressions of difference regarding the initiative being discussed versus mainstream or normative architectural education.

Finally, expanding the understanding of wood in architectural education requires 'multidisciplinary collaboration' – the participation of different stakeholders in the architectural and educational processes, both specialist (forest scientists, engineers, urban planners, and consultants, for example) and non-specialist (members of the public, for example). This informed a concept that considered collaboration across many disciplines. As we will discuss in Chapter 7, there are many different shades of collaboration, and these vary significantly in the different case studies.

The third step in the construction of our grounded theory saw these concepts grouped together into broad categories. In the case of the codes and concepts discussed here, these fell into a large and emergent category of education. These categories were produced not through interpretation but through correlation of common ideas and themes that were emerging from the research and which have become the armature of this book: namely Chapters 3–7.

The journey of constructing our own grounded theory of wood in architectural education from individual 'code' to collections of 'concepts' to broad 'categories' helped us to extract valuable, consistent information from subjective sources to create a grounded theory of architectural education. The five themes described above synthesise aspects of the use of wood in hands-on architectural education that help us frame and understand it in light of the challenges awaiting future architects and their teachers. This synthesis from codes to concepts to categories is achieved through alternating inductive and deductive coding: the interviews were an inductive tool that allowed interviewees to talk freely while remaining within a clear thematic framework; the analysis of the interviews has been a deductive process in which themes have emerged after an analysis of the results of the initial phase of coding.

In what follows, we present our findings in two ways: in short, descriptive case studies of the nine initiatives included in our study, and in longer more subjective chapters that explore our summative findings. Case study research is not without its critics, those who argue that one cannot generalize on the basis of just one or even a small number of cases. So there are two limitations to address: one that we argue is perceived and one that is recognised. Some perceive case studies to be subjective, both in their selection and their analysis, and therefore unreliable. We align ourselves with Bent Flyvberg, who argues instead that the case study method is both 'necessary and sufficient method for certain important research tasks in the social sciences, and it is a method that holds up well when compared to other methods in the gamut of social science research methodology' (Flyvbjerg, 2006, 241). Flyvberg refers to Thomas Kuhn in his defence of the benefits of case study research: any scientific discipline (in this case, architectural education) that cannot draw on a large number of high-quality case studies is a discipline without a means of systematic production of exemplars. A discipline with an absence of systematically produced case study exemplars is an ineffective one.

Secondly, there are recognised limitations to the use of qualitative grounded theory. The definitions of the codes derived from this work are a critical, partial, and subjective operation of the authors and cannot be generalised. The grounded theory method used to analyse a small number of interviews cannot be as rigorous as research involving hundreds or even thousands of respondents. In our specific case, the limited number of case studies cannot claim to be fully representative of the highly varied practices of design-build projects and live projects. Nevertheless, our hope is that this work can help to expand and diffuse the discourse on the topic of wood in architectural education, providing critically analysed information, synthesised through keywords, that could help teachers, students, and practitioners to evolve their methods towards sustainability.

Notes

1 Every Icelander – especially the taxi drivers who pick you up from the airport – will be delighted to tell you the country's most famous joke: 'What do you do if you get lost in a forest in Iceland? Stand up.'.

2 The complexity of the idiom, in British English at least, is deepened by the double-meaning of the word 'wood': wood is the name for the hard fibrous material that trees are made of, but also an area of land, smaller than a forest, in which trees grow. We seek to avoid any confusion about the double meaning in British English by exclusively using the word forest to describe the landscapes from which wood (and many other things) come.

Bibliography

Addison, Catherine. 2007. 'Terror, error or refuge: forests in Western literature.' *Alternation (Durban)*, 14(2): 116–136.

Brown, James Benedict. 2012. *A Critique of the Live Project*. Thesis (Ph.D.), Queen's University Belfast.

Brown, James Benedict, and Pete Russel. 2022. 'When design-build met the live project – or – what is a live-build project anyway?' In *Experiential Learning: Design-build and Live Projects in Architectural Education*, edited by Burak Pak and Aurelie De Smet. Abingdon: Routledge.

Chakrabarty, Dipesh. 2021. *The Climate of History in a Planetary Age*. Chicago: The University of Chicago Press.

Charmaz, Kathy. 2014. *Constructing Grounded Theory*. 2nd edition. London: SAGE Publications Ltd.

Cronon, William. 1995. *Uncommon Ground: Toward Reinventing Nature*. 1st edition. New York: W.W. Norton & Co.

Descartes, René. 1637. *Discours de la méthode*. Leyden: Ian Maire.

Flyvbjerg, Bent. 2006. 'Five Misunderstandings about Case-Study Research.' *Qualitative Inquiry*, 12(2): 219–245.

Ghosh, Amitav. 2016. *The Great Derangement: Climate Change and the Unthinkable*. Haryana: Penguin Books.

Harrison, Robert Pogue. 1992. *Forests: The Shadow of Civilization*. Chicago: University of Chicago Press.

Laing, R.D. 1967. *The Politics of Experience and the Bird of Paradise*. Harmondsworth: Penguin Books.

Mill, John Stuart. 1874. *Nature, the Utility of Religion, and Theism*. 2nd edition. London: Longmans, Green, Reader, and Dyer.

Shapiro, Sidney. 2001. *Jews in Old China: Studies by Chinese Scholars*. Expanded edition. New York: Hippocrene Books.

Thoreau, Henry David. 2014. *Walden on the Duty of Civil Disobedience: First Avenue Classics*. Minneapolis, MN: First Avenue Editions, a division of Lerner Publishing Group.

Ursprung, Philip. 2023. 'Touched by Wood.' In *Touch Wood: Material, Architecture, Future*, edited by Carla Ferrer, Thomas Hilderbrand, and Celina Martinez-Cañavate. Zurich: Lars Müller Publishers.

Case Study 1

Studio in the Woods, United Kingdom

DOI: 10.4324/9781003428930-3

Figure CS1.1 Studio in the Woods, 2011. Photo: Piers Taylor.

Our first case study is not situated in or affiliated with a university or higher education institution. It is an ephemeral and periodic workshop organised by a quartet of architects who were seeking an antidote to normative architectural education.

Studio in the Woods was established in 2005 by Piers Taylor (Invisible Studio), Gianni Botsford (Gianni Botsford Architects), Meredith Bowles (Mole Architects), and Kate Darby (Kate Darby Architects). The workshop is a short design-build studio for students in architecture and related disciplines, although participants have included artists, carpenters, and engineers. Several other invited practitioners join the core group of architects to lead four or five groups towards the realisation of a small, built project. This happens over the course of a summer weekend, usually from Friday afternoon to Sunday evening. Open to all, the workshop typically costs €300–400 including catering.

The diversity of the different groups' expertise and backgrounds generates different approaches to the project. In the initial phase of the studio, participants choose which group they want to join based on personal preference or attitude, in much the same way that students might vote or choose to join an elective design studio in a school of architecture. Together they then spend the next couple of days developing a project, typically using wood felled from the site.

The design activities are complemented by evening lectures from the guest practitioners. Participants camp on site, but eat together at mealtimes. The studio concludes with a walk through the realised projects in conversation with visiting critics, including Niall McLaughlin, Robert Mull, Peter Clegg, and Ted Cullinan.

Studio in the Woods originated in the UK, with instalments on the Isle of Wight and in the Wyre Forest, where it worked with a Community Land Trust as part of an exploration of alternative forest management strategies. In spite of disruptions due to the COVID-19 pandemic, workshops were proposed in 2021 and 2022 in France. Participating architects have included Barbara Kaucky and Susanne Tutsch (Erect Architecture), Je Ahn and Maria Smith (Studio Weave), Lee Ivett (Baxendale), Lynton Pepper (Architecture00), Shin Egashira (Architectural Association), Lena Gotmeh (Lena Gotmeh Architecture), Toby Lewis and Akos Juhasz (Feilden Clegg Bradley Studios), Fergus Feilden (Feilden Fowles), Guan Lee (Grymsdyke Farm and the Bartlett School of Architecture), Hannah Durham (Cullinan Studio), Charley Brentnall (Xylotek), Zoe Berman (Berman Studio), Carolina Vasilikou (University of Reading), Tim Lucas (Price and Myers), and Jack Hawker (Momentum).

Studio in the Woods grew out of a feeling, shared by the founding architects, that getting together outside of the confines of curriculum-based architectural education was important for the development of both teachers and students. The studio is a programmatic alternative to normative architectural education. It is affiliated with the Global Free Unit, a network of architecture studios and classrooms through which the student takes responsibility for their learning, which can take place outside of the framework of conventional academic institutions.

Working hands-on with wood is fundamental to this approach. Taylor reflects how, in normative architectural education, 'you never worked on anything that wasn't an approximation of the thing.' He continues: 'We make an approximation of the thing; we

Figure CS1.2 Gridshell, Studio in the Woods, 2017. Photo: Jim Stephenson.

work in a kind of code space which is either the digital realm or the physical realm of the model. And as architecture students, we never have the embodied sense of being in the space that we are imagining or helping us to imagine the space.' At the start of his career, Piers Taylor worked in Australia in the practice of the architect Glenn Murcutt. Taylor credits that experience with the formation of his interest in the criticality of the relationship between building and site. Taylor's practice, Invisible Studio, was established in a forest environment, an actual working woodland managed by the practice alongside the business of architecture.

Studio in the Woods offers participants (not just students, and not just students of architecture) an opportunity to engage in an intensive, fast, and 1:1-scale workshop, experimenting hands-on with small, built projects using locally produced wood typically harvested from the site in which they are working.

Figure CS1.3 Studio in the Woods, 2017. Photo: Piers Taylor.

Chapter 3

The Characteristics of Wood

This chapter explores how the properties and characteristics of wood make it a defining component of the initiatives studied in this book. What we find is that amongst all the many materials used in the construction of buildings, wood is an ideal material to be directly experienced in the education of architects. Why is this?

Perhaps it is because wood is *easy*. Even far from timber-producing regions, wood and timber are easy to procure. Once acquired, wood and manufactured timber components are usually easy to handle, easy to work with, easy to assemble, and easy to alter. The *easiness* of working with timber complements educational initiatives in which the timeframe is always constrained by the curriculum or calendar (and usually shorter than the time given to the construction of buildings), resources can be scarce, and the competence of participants is in continuous evolution.

Amongst the case studies presented in this book, the procurement of timber is frequently motivated by ease of access. Often this is due to geographical proximity with productive forests or developed wood economies.[1] Being situated in an area where timber is produced simplifies procurement and usually reduces costs. Standardised industrial timber products, including Canadian Lumber Standard (CLS) and structural timber, can provide a relatively low cost of structure when compared to masonry and steel.

DOI: 10.4324/9781003428930-4

Figure 3.1 In this pavilion constructed by participants of the Studio in the Woods summer school, plastic cable ties are used to quickly and easily bind together components of a wood grid shell structure. Photo: Jim Stephenson.

Except for the very largest and most structurally capable timbers or manufactured wood products, wood is easy to handle. Most lengths of sawn timber can be lifted and moved by two people of average strength, making it accessible for students taking their first steps on a building site. As a consequence, timber is also easy to assemble. A wooden frame can be rapidly composed with fixings made using simple handheld tools like drills and hammers. Joining timber elements using bolts, screws, and tails is tolerant to errors: accommodating an imprecisely cut piece or a non-perfectly matched joint can be done in a variety of ways.

In addition to all these characteristics, wood is a material with a complex but, as we will show, 'transparent' ecology. The process through which the material grows as part of a forest links it to a large number of natural, human, and non-human stakeholders.

Practising Wood in Architecture

The relatively short lifetime of a tree harvested for construction materials can make its ecology more easily graspable to students of architecture when compared to the hundreds of thousands of years necessary to generate non-renewable resources such as stone, cement, and steel. Understanding a process that lasts a few decades – and which can be connected to lived experiences of forest – makes it easier for a student to comprehend the invisible geological forces creating materials mined from below the earth's surface. For students of architecture, engaging with timber can be a means of immediately confronting the ecological processes that produced it, coping with its constraints while effectively channelling the forces it moves. We argue that this more accessible ecology can be defined as 'transparent,' meaning that it is more easily visible as a whole when compared to other ecological processes. Moreover, we imagine that this easier understanding of the ecological entanglements of a material can facilitate its sustainable use. As a consequence, we have taken the step of specifically not listing sustainability as a characteristic of wood since we do not assume that it is intrinsic to the material. Rather, sustainability can be a possible consequence of its other properties.

In the remainder of this chapter, we explore in more detail five characteristics of wood which we believe contribute to its suitability as a material to be experienced hands-on in architectural education.

3.1 PROXIMITY

The first characteristic of wood that we consider is its potential high degree of proximity. For many of our readers in the boreal regions of the planet, standardised sizes of timber can normally be procured locally, and often from local sources. Procuring locally produced timber has several positive implications. Firstly, using a locally produced material can contribute positively to the environmental and aesthetic integration of a new building in an existing context. If a material has been produced locally for a long time, there is a greater opportunity to exploit local expertise and to produce a building that compliments the existing built environment. Secondly, the use of locally produced materials can contribute positively to the local economy and society. Public universities, which make up the majority of the institutions hosting the

Figure 3.2 Freshly felled trees, Vännäs municipality, Sweden. Photo: James Benedict Brown.

initiatives described in our case studies, often have obligations to disperse their economic impact throughout the region where they are located. Thirdly, using a locally produced material can dramatically reduce the carbon emissions produced in the manufacture and transportation of materials.

It is important to note that local production is not the be-all and end-all. In several of our case studies, timber has been chosen where there is limited or no local production. In Mongolia, the GerHub community centre in Ulaanbaatar was built using imported materials, including Russian timber and Chinese channelised plastic. These materials may have carried a greater environmental impact than if they had been produced locally but were judged to be appropriate choices for the overall scheme, one which exploited passive solar gain to warm the building and reduced the amount of masonry and concrete needed. In South Africa, where the University of Nottingham's (UoN) Design+Build Studio has increasingly chosen to work with wood, timber is available but is not used to a particularly large extent. As a result, local sawmills and suppliers were keen to support a project that promoted the use of their material, which was often side-lined in favour of cement blocks. In these examples, balancing the economic and environmental costs of choosing an imported material is part of the holistic decision-making that informs the design of a building. For students, this is a valuable opportunity to understand and experience the calculations and decisions they will likely make in practice. Even when precise carbon or cost calculations are impossible to make, understanding how emissions produced by transporting building materials could equal or exceed the volume of carbon stored in the material used can provoke critical discussions about the true sustainability of using wood.

Proximity to materials has both practical and cultural consequences: ease of procurement is an obvious one, as being close to the production area facilitates search, transport, and overall logistics. In addition, being in close contact with the material source can be linked to a cultural familiarity with it, either pre-existing in the surrounding society or to be built within the educational or research initiative.

A critical and holistic engagement with a building material can bring a deeper understanding of its ecology. For example, at the Architectural Association's (AA) Hooke Park campus students work programmatically with the forest in a symbiotic way. Students work and study in the environment that provides the materials; they have the opportunity to live in a forest setting where the entire process from forest management to building takes place. This kind of proximity can enable a more intensive learning experience that gives the students a more complete overview of the implications of building with wood.

The Studio in the Woods workshop provides an even more intensive (three-day) learning experience deeply rooted in the site. The materials used for their projects come from the site itself and the resulting constructions relate to the local environment. The self-evident material continuity between the forest and construction contributes to a heightened sense of place in the work produced.

In China, the cultural value of the use of local materials is a crucial component in the strategy of the Insitu project. The research and, in some cases, the revival of traditional building techniques are deeply linked to the local community. Exploring the availability of locally produced materials is intended to re-establish a relationship between a landscape and its inhabitants. At the same time, realising buildings using local materials and traditional techniques – but updating them to produce modern building forms – is a way to rebuild social cohesion in close contact with the cultural and natural environment.

3.2 CARBON FOOTPRINT

The second characteristic of wood that we consider is its potential to have a low carbon footprint relative to other building materials. The use of timber and, in general, of biomass in buildings is considered to have a positive impact on the environment as they can work as carbon storage. During its lifetime, a tree absorbs carbon to grow. As long as the tree is not left to decompose or burned, the carbon that has been absorbed during

Figure 3.3 Woodchips piled up at Sävar sawmill near Umeå, Sweden. In addition to producing some 300,000 cubic metres of sawn timber every year, the sawmill burns waste material in three biofuel boilers to generate heat for both the plant and approximately 10,000 households in the neighbouring community. Photo: James Benedict Brown.

its lifetime will not be released into the atmosphere. Timber buildings can therefore be seen as a method of capturing and sequestering carbon that has been absorbed from the atmosphere, and which might ordinarily be returned to it in the event of decomposition due to natural die back or fire.

Comparing the production processes of other building materials, there is a striking contrast of carbon flows. Given its versatility and suitability for framed structures, steel is perhaps the most useful comparison. Steel is produced in an energy-intensive process that removes carbon from iron ore mined from the ground. The lower the volume of carbon in steel, the higher its structural performance. In the traditional production process, carbon removed from the iron ore would be released into the atmosphere as smoke. This, together with the emissions from the energy consumed, makes the production of steel one of the most carbon-intensive processes when compared to any other industry.[2] Whereas steel is most often used in larger buildings or those with structural requirements beyond the normal capacity of wood, smaller buildings are often built using blocks or bricks. The firing process of bricks and ceramic materials requires the sustained baking of clay at extremely high temperatures at a high energetic cost.

Unlike the manufacture of steel or masonry, the production of wood and timber products does not involve intensive energy consumption or the release of carbon. But the sequestration of carbon is not sufficient alone to ensure the sustainability of the forest and timber industries. These processes are linked inherently to the wider ecosystems of the forest, affecting a wide range of plant and animal systems. Different trees and plants, insects, mammals, birds, soil, waters, winds, and even human communities are all an integral part of an ecological system that needs to be complete in all its parts to actually be an effective carbon storage.

In the case studies presented in this book, the potentially positive carbon impact of wood products is often regarded as one of the basic characteristics of the material. The potentially low carbon footprint of wood products is a common motivation for its use, but it is also regarded as a critical design driver. However, amongst advocates for building in timber instead of other materials, there is also a strong caution against building with more wood than is absolutely necessary. In the projects of Stuttgart's Institute for Computational Design (ICD) and Institute of Building Structures and Structural Design (ITKE), timber buildings are not assessed according to their carbon storage potential, since increasing the use of wood could potentially have a negative impact on the forests producing it and the carbon-emitting logistical supply chain. On the contrary, ICD/ITKE's projects demonstrate the pursuit of material optimisation through advanced computational technologies and structural innovation, with a goal of using as little material as possible, thus lowering the overall impact of the building on the environment.

A similar principle is evident in the work at the AA's Hooke Park campus, where computational techniques are used to identify and use wood products that might be discarded by traditional forestry practices. Thinning, coppicing, and irregular tree sections, for example, are carefully parametrised with the help of scanners and software to assess their structural capacities and then use them as building elements. This aims to embrace the inherent properties of wood without recourse to standardisation.

In projects completed through the Aalto Wood Program in Finland, the overall environmental impact of timber is considered in both the raw material and the building life cycle. These projects aim to minimise the environmental footprint of the building by pursuing quality and, consequently, durability, thus extending the life of the building and making more efficient use of the raw materials.

3.3 VERSATILITY

The third characteristic of wood we consider is its versatility. Wood can be used as a construction material in many different ways. These include applications demonstrating different degrees of technological adaptation, from the assembly of raw elements harvested directly from the forest, through simple post and beam, balloon-frame, truss, cross-laminated timber (CLT), lamellar, or grid shell structures, to engineered timber. The great variety of wood products available allows for differing complexities of application, from simple and cheap structures to advanced structural systems that require computational tools to design.

Practising Wood in Architecture

Figure 3.4 Construction of Gradual Assemblies Pavilion, Istituto Svizzero, Rome, Gramazio Kohler Research, 2018. Photo: Gramazio Kohler Research, ETH Zurich.

This versatility makes timber particularly suited for use in a pedagogical setting where the availability of resources varies greatly between different contexts – for example, in terms of materials, technology, finance, or available hours. As demonstrated by the case studies presented in this book, different kinds of wood can adapt to and serve different pedagogical settings, from short design workshops to longer-term research and educational strategies.

Engineered timber can be exploited at a very advanced technological level. If educational initiatives have access to controlled environments in laboratories or workshops – for example, at IBOIS (the Laboratory for Timber Constructions at the École Polytechnique Fédérale de Lausanne [EPFL]) or Gramazio Kohler Research (GKR) – there is the possibility to experiment with and test innovative structural concepts. Pre-fabrication of complex elements before final construction allows for repeated practise and refinement. Technology can also be used to identify unforeseen potentials: at the AA's Hooke Park campus, the close availability of forest land allows experimentation with digital surveying of material for harvesting and optimisation.

These different approaches are often linked to specific research or pedagogic interests. In some of the case studies in this book, computational design is an area of interest, working towards a more efficient use of wood. In other case studies, the less technologically complex use of standardised wood elements responds to an interest in social sustainability or economic constraints, and the consequent need to optimise resources.

A more pedagogical interest is found in many of our case studies where the use of wood allows students with low levels of experience or training to profit from participation in a full construction process. This versatility of wood is crucial for initiatives of shorter durations, like the UoN's Design+Build Studio or the Insitu Project, in which students are afforded limited time on the building site and need to realise a building quickly.

3.4 ACCESSIBILITY

The fourth characteristic of wood we consider is its accessibility. In areas of the world with geographical proximity to productive forests or developed wood economies, procuring wood and timber products can be relatively straightforward. The ease of procurement of building materials is one of the contributing factors to the successful projects described in our case studies. Easy accessibility of wood allows focus to be turned to other aspects of its use in architecture, creating space for experimentation in material and design strategies. When wood is used at a relatively low level of technological complexity, procurement, handling and the necessary skills are simplified.

On the smallest scale and at the simplest dimension of accessibility, at the AA's Hooke Park campus and at the itinerant Studio in the Woods, the accessibility of local timber is embedded into the concept and mission of each initiative. Both harvest timber on site. This is a great facilitation for the procurement process that minimises dependency on third-party providers and logistic networks, thus also enhancing the sustainability of their activities. This local accessibility is what allows them to directly experiment on site and to give students a complete overview of the process behind the realisation of a timber building.

Several of the case studies in this book are based in productive forest regions. The ICD/ITKE research centre is, for example, located at the core of a region where forestry industry is of great importance. This gives the institution a strategic role in promoting advanced research on timber. In a similar way, Gramazio Kohler Research and IBOIS are located in the trans-national Alpine Region where forestry and the timber industry contribute to a developed local wood economy. In Finland, the Aalto Wood Program is part of a wider economic and political environment in which forestry, timber industry, and timber buildings play a crucial role. The Finnish state is actively pursuing a strategy to promote the use of wood in the building industry and Aalto Wood Program is indirectly part of this national strategy.

3.5 TRANSPARENT ECOLOGY

The four characteristics of wood described in the preceding sections lead us to an overarching characteristic that we argue is key to understanding the potential of wood for sustainable construction: its transparent ecology.

An ecology is a web of ever-changing relations between parts. To act sustainably in relation to any ecological system is to preserve a balance which does not extract more than the system can regenerate. A consciousness of the complexity of an ecological system is necessary in order to act upon it in a sustainable way. Timothy Morton defines this kind of ecological thought as 'thinking big' (Morton, 2010). Ecologies are interconnected systems and a change in any single factor can influence a number of others. In northern Sweden, for example, clear-cutting of forests and replanting with monocultural plantations has been found to dramatically reduce hanging lichen populations, thereby limiting the availability of food for reindeer in the winter months. Thinking ecologically means not only caring about the sustainability of a single material but also being able to comprehend the complexity of an ecological system.

Wood is very often labelled a renewable material. Mineral or fossil materials, which take millions of years to be formed, cannot be renewed as quickly as they are harvested. The extraction of mineral and fossil materials from quarries, mines, or oil fields is often pictured in opposition to a self-evidently renewable regeneration of forests. A forest can, with the right approach, be managed in ways that allow for the regeneration of the trees that are harvested. A much greater attention to ecological complexity is required if plant and animal life is not to be affected negatively. Harvesting wood affects the complex ecological system of the forest. But it also changes other systems: it shapes and re-shapes vast landscapes. Wood is handled and processed by complex economies, affecting human and non-human stakeholders alike (Ibañez, Hutton, and Moe, 2019).

Setting aside the legitimate interrogation of the sustainability of modern forestry, the almost unimaginably long timespan required to create mineral and fossil materials makes it harder to grasp the enormous forces that generated them and the complexity of their ecological interactions. The ecological systems from which wood is harvested are easier to observe and, therefore, to understand. As a result, wood is a good pedagogical tool in architectural education: the possibility of fully understanding the complex ecology behind a material can teach future architects not just to choose supposedly sustainable materials but to actually understand the complex consequences of their design choices. This *transparent ecology* is what we believe is essential to achieving a truly sustainable architecture: the ability to comprehend the complexity of the environment from which a material is harvested.

Whether it is in the rough, unfinished walls of a log cabin or the unfinished cross-laminated wall panels of the Wood Hotel in Skellefteå, Sweden, wood is a material that provokes many positive emotional reactions in humans. Like all materials, wood changes over time as a result of exposure to the atmosphere and usage, demonstrating an active relationship with the surrounding environment. Much of the popular hostility to concrete, which defined a large swathe of twentieth-century modernism, can be ascribed to the negative perceptions of concrete's tactility and ageing.

In this book, we see how architectural educators have recognised timber's capacity to show its ecological implications and how this has affected their pedagogical approaches. At the graduate level, the ICD in Stuttgart supports students to connect first-hand observations of the microscopic structure of wood with a deeper understanding of its application as a building material. At the AA's woodland campus in Hooke Park, students can engage in the full process of harvesting, selecting, processing, testing, and fabricating wood. A holistic connection between forest and final product engages students in a far deeper ecological understanding of the complexity of wood as a potentially sustainable building material. And, finally, in design-build buildings realised by the Insitu Project in China, students are able to witness how contemporary applications of wood connect the material to shared cultural roots.

To build with wood is to build with a product of the environment in which we live. If we are prepared to treat wood as the product of complex ecological systems, it provides us with a unique opportunity to perceive those ecologies. By engaging with older techniques for wood construction, we see how wood buildings have always been rooted in the complex ecological systems that produced them, establishing aesthetic relationships between buildings and environments. As wood is defined, specified, and traded as a commodity, it becomes ever more important for students of architecture to be made aware of these connections.

Notes

1 An example of a developed wood economy is one in which locally or domestically produced wood is complimented by a diverse and reliable supply of imported products. In the United Kingdom in 2022, for example, approximately ten million tonnes of softwood and hardwood were harvested, and some six million tonnes of sawn wood was imported. See https://www.forestresearch.gov.uk/tools-and-resources/statistics/statistics-by-topic/timber-statistics/uk-wood-production-and-trade-provisional-figures/.

2 Work is ongoing to address the high carbon emissions of steel production. SSAB AB, formerly Svenskt Stål AB, whose largest shareholders are the governments of Sweden and Finland, delivered the first batches of steel manufactured without the use of fossil fuels in 2021. The so-called 'fossil-free steel' is manufactured using hydrogen instead of coking.

Bibliography

Ibañez, Daniel, Jane Elizabeth Hutton, and Kiel Moe. 2019. *Wood Urbanism: From the Molecular to the Territorial*. New York; Barcelona: Actar.

Morton, Timothy. 2010. *The Ecological Thought*. Cambridge, MA: Harvard University Press.

Case Study 2

Hooke Park, Architectural Association, United Kingdom

DOI: 10.4324/9781003428930-5

Figure CS2.1 Research Station at Hooke Park. Photo: Postgraduate Design + Make course, Architectural Association.

Hooke Park in Dorset, England, is a satellite campus of London's Architectural Association (AA), located about 200 km south-west of the capital city. The park consists of 330 acres of forest and a small forest campus for teaching, forest management, and manufacturing. Hooke Park operates as a productive forest, uniting education, research, forestry, and wood fabrication.

The campus was established by the furniture designer and manufacturer John Makepeace (b. 1939) in the nineteen eighties as the School for Woodland Industries. Makepeace had formerly established the Parnham Trust and the School for Craftsmen in Wood, later known as Parnham College. The School for Woodland Industries sought to integrate the teaching of furniture design and manufacture with forestry. The first building on the campus was completed in 1986 according to a design by Frei Otto. Otto designed a building with a tensile roof that used spruce thinnings sourced from the surrounding forest. After the campus was transferred to the AA, several other buildings were realised, making use of innovative, readily available timber and digital technologies.

Figure CS2.2 Pre-fabrication of frame for the Wakeford Hall Library inside the Big Shed at Hooke Park. Photo: Postgraduate Design + Make course, Architectural Association.

The AA's Design + Make graduate Master programme (MArch/MSc) is based at Hooke Park, conducting teaching and research through 1:1 fabrication.

Supported by a diverse team of expert practitioners, students of the MArch/MSc programme inhabit an environment that combines forest, studio, workshop, and building site. The campus is also open to other students at AA for shorter stays. Hooke Park's large-scale fabrication facilities act as a testing ground where they can develop hands-on research through prototyping. The Design + Make programme itself focuses on the continued expansion of the Hooke Park Campus. Students are actively involved in the realisation of buildings for the campus using materials sourced from the surrounding forest. MArch students use full-scale building projects on the Hooke Park campus as a vehicle for design research. Formulating individual research interests within a group project, each student subsequently reflects critically on aspects of the work in their written thesis. MSc students, meanwhile, are encouraged to adopt a more explicit technological focus on the innovative application of timber in architecture, which is developed and tested through full-scale system prototypes using new fabrication technologies.

Students embarking on the graduate programme at Hooke Park, who typically come from more conventional architectural education pathways, must engage with a design method based on making and in which digital tools are just an instrument. This can be unsettling for them, as it poses the design as an outcome more than the goal of architecture practice. Students engage with the material and are invited to conceive design solutions basing on its behaviour. At the end of the course, students tend to have developed confidence with both hands-on and advanced computational wood work.

Research undertaken at Hooke Park seeks to advance the materialisation of architecture through the synthesis of advanced technologies, craft techniques, and a deep understanding of natural materials. A vital proposition is that new digital design and fabrication technologies can enable traditional manufacturing techniques to be re-invented. Emerging digital tools such as digital three-dimensional scanning of trees and forests, generative modelling, and robotic fabrication provide new opportunities for replicating the feedback between natural geometry, material properties, and designed form that had previously connected the designer, maker, and artefact.

Hooke Park is a residential campus, with student accommodation provided in a number of small lodges and cottages. There, students, teachers, and researchers live in close contact with the industrial landscape of the forest, witnessing different stages of the life cycle of trees and observing the processes of harvesting wood to build with and maintaining the forest landscape. The intention of the programme is for students to develop a deep understanding of the materials and their peculiarities. This is the premise for ongoing research in the use of supposedly 'zero-value' materials discarded by traditional industrial forestry. This includes the use of thinnings, coppiced or pruned wood, and other small-diameter pieces.

The pedagogical approach to teaching at Hooke Park lies in the inversion of the normative relationship of the architect to the

Figure CS2.3 Designing and constructing around the irregular shapes of trees felled from the surrounding woodland. Photo: Postgraduate Design + Make course, Architectural Association.

forest, the source of the materials they specify. Instead of demanding a material from the forest according to pre-determined standards or dimensions, the architect is conceived of as an individual with the tools and capacity to see what the forest can provide (see Chapter 6 for an example of this in the Wood Chip Barn). A more symbiotic relationship between architecture, forest, and wood is envisioned as contributing to a more circular pattern of production and consumption.

The design of each of the campus buildings at Hooke Park is derived from the characteristics of the wood used in their construction. This approach recalls historic ship-building industries of the nearby south coast of England, whose experts surveyed forests to identify components of ships from appropriately shaped trees. By engaging with an organic architectural language of natural geometries found in the forest, as opposed to manmade geometries, these projects seek to reduce the 50–60 percent of wood mass that is estimated to go into waste streams in the

Figure CS2.4 Research Station at Hooke Park. Photo: Postgraduate Design + Make course, Architectural Association.

production of standardised building components by industrial forestry.

One of the overarching goals of recent activities at Hooke Park has been to reduce the amount of waste produced by forestry activities and timber production. What is commonly considered zero-value material is at the core of the initiative. Of all the case studies in this book, Hooke Park is perhaps the one with the lowest apparent interest in the social impact of architecture. However, the programme demonstrates an intention to train architects who interact with their environment as a stakeholder rather than a consumer. Through wood, they try to act as a stakeholder with an ethical position derived from their being an educational institution.

Chapter 4

Designing and Building with Wood

This chapter explores ways in which the use of wood affects the process of design and building. It shows how wood can stimulate research into innovative design solutions, encourage a more resource-conscious approach to building, affect the understanding of the building process, and promote consideration of the building life cycle during design and construction phases.

The case studies in this book offer a diverse range of examples in which the implications of building in wood are explored with different agendas. These include, to give just a few examples: attempting to extend the physical properties of wood; exploring how to realise a community building as fast and cheaply as possible; and updating traditional building techniques to modern forms.

Wood presents unique technical challenges compared to fossil- or clay-derived materials. Its sensitivity to external environmental conditions, the different characteristics of the various species, its harvesting processes, and the difficulty of reconducting it to industrial standards all introduce elements of uncertainty. Industrial production processes often result in substantial amounts of waste. Obtaining dimensionally homogeneous cuts from round tree trunks or using preservative chemicals produced with a significant environmental impact are just two examples of this. These methods can significantly worsen the environmental

DOI: 10.4324/9781003428930-6

performance of wood, bringing into question the true sustainability of the material.

The case studies we analyse in this book address these uncertain characteristics of wood not as issues to be controlled but as occasions for rethinking our approach to the material. Through innovative design approaches that embrace and explore the natural properties of wood, they challenge the conventional, extractive approaches that often characterise the current timber industry. They present a broad spectrum of technological and design solutions aimed at promoting more sustainable architecture. Ultimately, these approaches can provide valuable insights for students on how to interpret architectural practice responsibly.

4.1 DESIGN INNOVATION

The choice to build with a specific material is not just a matter of aesthetics: it also implies technical and structural constraints that inform every aspect of the design.

The many ways in which wood and timber can be used in construction are all characterised by a certain range of spatial

Figure 4.1 Construction of the Gradual Assemblies Pavilion, Istituto Svizzero, Rome, Gramazio Kohler Research, 2018. Photo: Gramazio Kohler Research, ETH Zurich.

Practising Wood in Architecture

possibilities, linked to the ways in which different elements are assembled and work as a structure. We refer to this as the inherent versatility of wood. We might, in another book and another time, have argued about the inherent versatility of concrete or steel. But wood offers such a direct connection between the environment that produces it, the raw materials, and the processes of construction, that wood is versatile differently and more transparently.

In the case studies presented in this book, the technical possibilities and physical constraints of wood and timber have been used as opportunities to experiment with spatial design. The architects, teachers, and students involved in these case studies have experimented either by rethinking traditional techniques to create modern, sustainable spaces or by expanding the technical potential of wood through advanced research. The spatial outcomes of these projects is, therefore, a direct consequence of their material focus.

At the Institute for Computational Design/Institute of Building Structures and Structural Design (ICD/ITKE) in Stuttgart, a powerful understanding and application of computational design and fabrication allows participants to experiment with innovative structural concepts. ICD/ITKE's research on shell structures and multistorey timber buildings directly relates to different possible spatial configurations: in the former by creating innovative forms, and in the latter by allowing for greater freedom of architectural space, thus expanding the possibilities of wooden structures and making them a viable alternative to more conventional structures.

At the Architectural Association's (AA) Design + Make campus at Hooke Park, the use of non-standardised timber elements directly affects the appearance and the spatiality of the buildings they realise. In their experiments, timber is not adapted to the architectural concept. On the contrary, it is the projects that embrace what the available timber affords. This includes aspects related to the form of the single elements, which is preserved as much as possible, but also to the actual availability of the material that is sourced on site from the local forest. By bringing together computational technologies and direct care of forestry, they manage to establish a sustainable building cycle that adapts to the life cycle of the forest instead of bending it to the needs of industrial standardisation. What is built is not what a designer has decided to realise but what the forest can ecologically sustain. The campus buildings, realised with the contributions of renowned architects such as Ted Cullinan and Frei Otto, reflect this philosophical and pedagogical approach to using wood.

Something similar happens on a much smaller scale at the itinerant Studio in the Woods, where in the short timeframe of a long-weekend workshop, participants realise small projects using timber elements directly sourced from the surroundings. This directs students to adapt their design to the material, reversing the normative educational paradigm in which space configuration prevails on material capacities.

At IBOIS (the Laboratory for Timber Constructions at the École Polytechnique Fédérale de Lausanne [EPFL]), the goal of eliminating non-wooden components in timber assembly pushes students and academics to conceive of innovative structural concepts based on the actual capacities of the material at hand, generating innovative forms and spaces. Their design process is based on an understanding of the final form of the building as an envelope: this implies conceiving structure, material, form, and hygroscopicity (its ability to absorb moisture from the air) as interdependent aspects of the design in which every choice influences every aspect of the building's performance. Therefore, the form is not imposed on the material but is consequential to its technical characteristics.

The design of buildings delivered by the Insitu Project in China is characterised by the rediscovery of traditional construction techniques and updating them to modern needs, identities, and material availability. The design of these buildings promotes the use of traditional building techniques by readapting them to more modern forms. One of the goals of the Insitu Project is to rekindle local knowledge about materials and use it to realise projects that can reinforce the local culture and identity. Local materials and updated traditional techniques can contribute to realising buildings that integrate more harmoniously into the local environment and at the same time restore an ecological relationship with wood. Traditional building techniques were deeply linked to the locally available materials and they developed solutions adapted to their characteristics. By promoting their rediscovery, the Insitu Project aims to re-establish an ecological relationship between rural communities and their environment. Designing with wood then becomes not a simple material choice but a cultural strategy that reverberates through the form of the resulting buildings.

Consciously designing for either wood or timber – by engaging critically with the types of materials that are available – can allow the design process to discover or apply innovative approaches. Considering the qualities of the material at hand allows for specific architectural solutions that can better exploit these characteristics. This approach to design is a reversal of the normative process of designing a building first and then specifying the materials second. It forces the architect to engage more consciously and ecologically with the material, based on a deeper understanding of its technical properties, availability, and affordability.

Figure 4.2 Paper models testing roof forms for the Kore Modular House, Wood Program, Aalto University. Photo: Alexander Garduno.

4.2 BUILDING ECONOMY

In Chapter 3 we explored how timber can be easy to procure and easy to handle and in the preceding section we have considered how wood and timber can be used in a variety of different ways. This wide range of applications can be representative of the many levels of complexity at which wood can be used. Manufactured timber products can achieve incredible structural and aesthetic feats, spanning wide distances and creating breathtaking structures. But wood and timber can also be used at a relatively simple level of complexity and be put in place quickly without a need for specialised or experienced workers. This makes it ideal for projects with low budgets and short time spans for realisation, contributing positively to the economic bottom line.

More advanced experiments with timber building explore ways to use it more efficiently, realising more lightweight structures but also trying to reduce waste by using less material. Not only is the economy of the project positively affected but also the amount of material is reduced, lessening the impact of a building on the environments that produce its wood. Among the research-active case studies we explore in this book, initiatives like IBOIS, ICD/ITKE, Gramazio Kohler, and Hooke Park all share a common focus on the optimisation of the material. They employ computational technologies and manufacturing to explore in depth the properties of timber, in order to better exploit them. This is achieved by reducing the need for other materials, mapping the structural properties of non-standardised elements, and experimenting with new, more efficient structural concepts.

At the other end of the scale of material and structural complexity, the projects realised in South Africa by the Design+Build Studio at the University of Nottingham (UoN) use standardised timber components for economic and programmatic effectiveness. Here, the economic aspect is crucial, since the buildings are realised using the funds raised by students and donors. Their buildings need to be finished in a very limited amount of time and to the best possible quality. Timber is ideally suited to achieve this. It can be procured easily and handled without the need for cranes or mechanical assistance. Even the use of pre-fabricated roof trusses in the Lesedi Creche (2019) did not prevent students from lifting the components on to the building by hand. Students can work with standardised timber after little training, reducing the need for specialist consultants, workers, or machinery.

In such contexts, it is important to consider skill as a resource of equal importance to materials. The Insitu Project in China does just this. When using local materials and techniques, relying on local knowledge is a way to enhance the effectiveness and economy of a project, through an understanding of sustainability that encompasses environment, economy, culture, and society. Local knowledge reduces costs but also builds on local culture, contributing positively to sustainable local economies.

4.3 BUILDING PROCESS

Designing using wood also has an influence on the building process. The peculiarity of the material and the ways it can be manufactured and assembled have direct consequences on the building site workflow. In projects driven by a tight budget, using wood can simplify construction while also adding flexibility and tolerance on site. In projects driven by material innovation, the evolution of wood technology and fabrication also produces changes in the building process, questioning the traditional sequence of designing preceding construction.

Technologies like computational design and robotic manufacture allow for a renewed building process in which fabrication becomes crucial. Consequently, the design and building phases are no longer two separate steps, but part of a single, continuous process.

Digital fabrication can enable the building process to become more dynamic and adaptable, allowing for modifications during the process without disruption. The possibility to produce tailor-made elements that can be assembled on site can further simplify and accelerate the building process, with additional advantages like reduced production of waste. Amongst our case studies, ICD/ITKE promotes digital fabrication not only as a way to optimise materials use but also to promote the integration of the two phases of design and fabrication, improving control over the building process and thus allowing for a higher degree of flexibility.

The projects undertaken at IBOIS in Lausanne strategically aim to involve local industrial and administrative stakeholders in order to establish a local, integrated, and sustainable building process based on careful forestry and the full implementation of computational technologies. At Gramazio Kohler Research (GKR) at ETH Zürich, meanwhile, a critical approach to manufactured timber such as glulam or cross-laminated materials seeks to reverse the attempt to make wood behave like steel or concrete. Here, the building process begins by using computational control and robotic fabrication to embrace and work with the naturally occurring strengths and weaknesses of each component. Likewise, at the AA's Design + Make programme at Hooke Park, projects explore the integration of forestry, design, and the building process, promoting constructional techniques that take into account the ecology of the forest and adapt around it.

Simplifying the building process of a project can also facilitate the involvement of a wider number of people in the construction. Unskilled or inexperienced stakeholders in a project, whether they are undergraduate students or members of the community in which a building will be delivered, can be quickly taught the simple techniques needed to handle and construct timber buildings. Creating opportunities for students and community

Figure 4.3 Using standardised materials and basic power tools facilitates construction by students. Photo: Thabang Radebe.

Practising Wood in Architecture

members to work alongside one another can lead to the mutual exchange of knowledge and experience.

The UoN's Design+Build Studio engages with projects where timely delivery is of crucial importance. Because students are engaged in delivering childcare buildings in neighbourhoods where there is an urgent need for such facilities, completing their work within the tight timeframe of the build becomes a moral obligation. They have to deliver a complete, efficient, and durable building in a very short timeframe. Using timber is one of the strategies employed to achieve this goal. Standardised manufactured timber products are used to assemble timber frame buildings with simple bolted connections that can be measured, drilled, and assembled by students with little training, thus increasing the number of people that can materially contribute to the project and its completion. The buildings they deliver aim to propose a modern, decent, and sincere architectural aesthetic (without resorting to cheap corrugated sheet metal or recycled waste materials, often favoured by students from outside Africa) to create a positive stakeholder perception of building quality.

Figure 4.4 Miaoxia Community Guesthouse under construction, 2018. Photo: Insitu Project.

In the Insitu project the building process is intended as a mutual learning exercise in which students and designers exchange knowledge with the local community. Consequently, this challenges and revitalises traditional building techniques and restores the relationship between the community and the local environment from which the material is sourced.

4.4 DESIGNING FOR BUILDABILITY AND CIRCULARITY

Normative architectural education tends to encourage students to develop their spatial configurations first and then choose materials and structures second. Most normative higher education institutions separate instruction in the fields of design, technology, structures, and construction into different courses, making it difficult for effective synthesis, especially when these subjects are taught in abstraction. But in all the design-build initiatives we explore in this book, we find alternative approaches to material choice and spatial configuration. They give opportunities for students to confront the realities of both design and construction.

In the 1990s, scholars in the field of construction management coined the phrase 'buildability' to critique the ways in which the designers of buildings were not contributing to their efficient construction (CIRIA, 1983). The isolation of designers from the construction process was identified as the cause of this lack of efficiency, something that Miller ascribed to the inherent dichotomy between the master builder (and subsequently the architect) and the various trades involved in constructing a building (1990). Richard Hyde writes that:

> ...the concept of buildability is not seen as an absolute quality or goal ... Categorizing buildability into good and bad methods of buildability seems to deny the complexity of the building process where factors that control the way buildings are built are closely interrelated, and also that buildability is related as well as relative to a range of qualitative aspects of buildings.
>
> (1995, p. 55)

Hyde offers a useful definition of buildability that is related to a number of contributing factors: production efficiency (the speed of erection, waste reduction, communication efficiency); the interfaces between elements (craftsmanship, building tolerance); the aesthetic expression (ordering of parts, joint articulation, visual grammar of construction, logical expression); ease of replacement, refit, or maintenance; and building simplicity (minimizing limitations of materials, double function of parts, detail simplification, optimizations of labour and materials).

In the preceding sections of this chapter, we have seen a variety of different ways not only of designing for buildability but of introducing these approaches to students of architecture. Looking back at Hyde's explanation of the concept of buildability for built environment professionals, we can see how these same contributing factors can be introduced to students when they work hands-on with wood. They get the opportunity to see how an understanding of a material's characteristics supports efficient production, minimizing waste or enabling quick construction. They get the hands-on opportunity to design, prototype, and then assemble interfaces between timber components, from complex joints that have been designed in the computer to simple nut-and-bolt right-angled junctions. They also get to see how wood allows for strong, logical, yet elegant aesthetic expression, and often in a way that facilitates the ease of maintenance or adaptation later.

Architecture and construction tend to function as part of a linear economy in which raw materials are used to make buildings that, if not repurposed or reused, will be turned into waste at the end of their useful life. Architects are increasingly adopting approaches to making buildings that are more restorative and regenerative. Designing for circularity challenges the architect to think about the reuse, repair, refurbishment and recycling of building components. Wood does not have to be disposed of when a building is demolished, but this is only possible if decisions made at the design stage plan for circularity. The use of metal fixings and fastenings or adhesives can condemn useable wood to landfill or incineration if they cannot be removed. Designing for circularity starts at the beginning of the design process, and connecting the acts of design and building in wood – especially in education – can help train future architects for sustainable practice.

Bibliography

CIRIA. 1983. *Buildability: An Assessment.* London: Construction Industry Research and Information Association.

Hyde, Richard. 1995. 'Buildability as a design concept for architects: A case study of laboratory buildings.' *Engineering, Construction and Architectural Management,* 2(1): 45–56. https://doi.org/10.1108/eb021002

Miller, G. 1990. 'Buildability – A Design Problem.' *EXEDRA,* 2(2): 34–38.

Case Study 3

The Institute for Computational Design (ICD) and the Institute for Building Structures and Structural Design (ITKE), University of Stuttgart, Germany

DOI: 10.4324/9781003428930-7

Figure CS3.1 The Buga Wood Pavilion, 2019. Photo: ICD/ITKE University of Stuttgart.

The Institute for Computational Design (ICD) and the Institute of Building Structures and Structural Design (ITKE) are two research centres at the University of Stuttgart, Germany that participate together in the Cluster of Excellence Integrative Computational Design and Construction for Architecture (intCDC). ICD focuses on computational design and the development of computer-aided manufacturing processes in architecture, whilst the neighbouring ITKE works on structural engineering and its integration with architecture.

As a joint research centre, they perform research on different structural materials like wood, fibres, and aggregates, working with innovative technologies and approaches like computational design and robotisation. Together they coordinate the MSc programme Integrative Technologies and Architectural Design Research (ITECH).

ICD/ITKE research innovative structural principles using wood, focusing mainly on two research trajectories: segmented shells and multistorey buildings. Their main goal is minimizing material use by expanding its performance through the exploitation of its inherent physical capacities. Wood is important to this research for two main reasons: firstly, its workability and machinability make it

Figure CS3.2 Researchers and students working with two industrial robots to pre-fabricate a component design in the ICD Studio, 2019/2020. Photo: R. Faulkner and ICD/ITKE University of Stuttgart.

an ideal material when it comes to the computational fabrication and additive manufacturing processes being researched. Secondly, in a world where population and urbanisation are expected to grow, research in material-efficient and low-carbon building structures is crucial to promote sustainable urban growth.

In fact, through using advanced computational tools and computer-aided manufacturing (CAM), they experiment on turning the natural hygroscopic behaviours of timber, traditionally considered as an obstacle to its standardised industrial use, into advantageous characteristics that can be exploited to obtain a higher structural, and therefore material, efficiency. Computation allows them to explore the properties and behaviours of wood from the molecular scale and to effectively exploit them, to the point of programming their structural behaviour and response to external solicitations. This approach brings together structural and spatial research, making it an inherently multidisciplinary work where the collaboration between architects and engineers, but also other disciplines like biology, is crucial.

One of the main goals of this research is to expand the spatial possibilities of timber structures. As traditional timber building does not allow for great structural spans, limiting the variety of spatial configurations, their work on single-storey shell structures and multistorey timber building is aimed at expanding the spatial possibilities of these structures, allowing for a higher spatial freedom.

As mentioned before, the pursuit of material and structural efficiency is also an environmental strategy aimed at minimising the use of the material. This implies that the use of structural timber as carbon storage, one of the dominant advocacy points in the global discourse around the sustainability of timber, is not considered by them to be an effective sustainability strategy. Their goal is instead efficiency of material use, and the consequence is buildings that are characteristically lightweight.

ICD/ITKE's geographical location is also crucial in the success and continuity of these research initiatives. Situated in south-west Germany, the two institutes benefit from a wider region that includes the Alpine areas of Austria and Switzerland where forestry is well established and industrial production is substantial. This makes their work strategic for the local economy and allows them to establish effective collaborations with industrial partners on projects with both research and educational aims.

In collaboration with the Institute of Building Structures (IBK2), ICD and ITKE engage in education, co-managing the MSc in ITECH. This program engages students in the realisation of experimental pavilions through which they test innovative structural concepts. The aim of the program is to train research-oriented practitioners who will be able to explore and rethink building practices using computational design, simulation, and fabrication.

Through the years, ICD and ITKE have tested several structural principles through the realisation of design-build projects. In their first phase, the projects were realised within the timeframe of the design studio with internal funding, allowing the students to follow the entire project. Subsequently, the success and growing reputation of their work generated an interest from the industry that led to collaborations and proper building commissions.

Experimental pavilions built with the support of students showcase the research undertaken at the two institutes and the new possibilities of these innovative building methods. The BUGA Wood Pavilion (2019) in Heilbronn used a segmented shell superstructure. It applied the structural principle found in the shells of sea urchins, executed with the use of experimental joints, adhesives, and robotised manufacturing. The Urbach Tower (2019) in Urbach is made of self-shaped, curved timber panels, realised by exploiting and programming the natural capacity of timber to change its shape when its internal moisture content decreases. This avoids energy-intensive transformation techniques, producing high-performing, sustainable building elements. Students involved in these projects are exposed to the possibility of continuing towards further research in the field of computational design. During their studies, they are confronted with the inherent structural qualities of wood and experiment with new technologies to exploit them. In this way, they can understand that the complexity of timber, rather than something to be normalised, allows it to be used more effectively and, therefore, sustainably. While other materials are used, Luis Orozco and Anna Krtschil explain that 'the reason we choose to use timber for both these rather different research trajectories is because it is an easy material to machine and to work with'.

Figure CS3.3 Urbach Tower, 2019. Photo: ICD/ITKE University of Stuttgart.

Case Study 4

Laboratory for Timber Construction (IBOIS), EPFL Lausanne, Switzerland

DOI: 10.4324/9781003428930-8

The Laboratory for Timber Construction (IBOIS) at the École Polytechnique Féderale de Lausanne (EPFL), Switzerland is a research centre that focuses on innovative structural concepts using wood. Since 2004, the institute has been directed by Yves Weinand, who coordinates an interdisciplinary group of architects, engineers, mathematicians, and computer scientists.

The laboratory aims to develop innovative timber building techniques and sustainable building processes. Laboratory members work at the intersection of architecture and structural engineering, focusing on the link between material choice, fabrication, and spatial outcomes. Major strategies address material efficiency and waste reduction through research into structural approaches such as timber rib shells, folded timber plate structures, woven timber structures, integral mechanical wood–wood connections, and robotically assembled timber structures.

Design is treated as an integrated process in which forest management, material procurement, design, and fabrication are all interconnected. This is supported by dialogue with industrial stakeholders in forestry, material production, local communities,

Figure CS4.1 Roundwood joint prototype. Photo: IBOIS.

Case Study 4: IBOIS

and politicians. IBOIS has established relationships and collaborations with several local stakeholders, promoting new building processes that have been implemented by industries and administrations.

One of the main foci at IBOIS is the reduction of waste and additional (potentially polluting) materials in timber construction. This has led to investigations into wood–wood joinery, reducing the use of adhesives in timber production and the need for steel fastenings. Computational technologies are also used to experiment with the use of round timber elements, in an attempt to minimise the biomass waste that results from planning and cutting tree trunks into more standardised rectangular section elements.

Although IBOIS conducts research, it also delivers a range of teaching, from short to semester-long courses and the supervision of year-long Master's projects. In Studio Weinand, one of the architecture programme's elective units, students are challenged to conceive new timber structural solutions that address the whole logistical process of timber production and fabrication. This studio takes place during the fifth and sixth semesters of the

Figure CS4.2 Roundwood joint prototype. Photo: IBOIS.

Figure CS4.3 Trade fair competition entry, 2018. Photo: Yves Weinand Architectes.

Bachelor course and first and second semesters of the Master's. The studio teaches future architects to understand that selecting a material for a building project is an aesthetic, technical, and *civic* act, having consequences for the building and the environment, but also for society. Students learn to experiment with parametric design and to test the ease of fabrication of their concepts with the help of 5-axis CNC machines. The underlying idea is that architecture, structure, form, and material should be considered holistically to achieve sustainable innovation. Through this approach, students are taught that a project is not only the result of a creative effort but rather the creative synthesis of a complex series of environmental and human factors: the available resources, the available technical capabilities, and the constraints of fabrication should be put together to produce innovative and sustainable spaces. Although Studio Weinand is not the only elective design studio at EPFL to offer students the chance to produce a physical output, IBOIS' teachers believe that Studio Weinand has become popular thanks to a growing interest amongst students of architecture in the craftsmanship and production of buildings.

Weinand explains how 'we still believe that construction methods and logistics are part of the architectural design and that we need to change the architectural design [process] … by reinventing or inventing new construction methods.'

Case Study 4: IBOIS

Chapter 5

Teaching with Wood

In the preceding chapters, we have explored the unique characteristics of wood and their impact on architectural design practices. But how do these characteristics come into play when wood is used as a pedagogical tool in architectural education? This chapter explores the reasons and potential for using timber in architectural education. With the building industry accounting for a significant portion of global greenhouse gas emissions, architectural education needs to provide future practitioners with the technical and critical tools to reduce the environmental impact of their work. This calls for a rethinking of pedagogical methods and goals, making sustainability not only a collateral discipline within a study program but an overarching critical point of view that needs to be at the basis of any design action. The potential sustainability of wood as a building material makes it an ideal tool for this kind of pedagogy. In this chapter, we consider how the ease of use, affordability, and versatility of this material can facilitate the implementation of hands-on architecture teaching, making it easier to work within the strict timeframes and economic constraints of an educational setting (Anderson, 2019).

Hands-on architectural education initiatives are characterised by their critical position towards established methods in architecture pedagogy and their detachment from real-world practice. By facilitating the implementation of hands-on pedagogies, wood can be seen as an element of this critique, exposing students not only to its material characteristics but also to the complexity of its ecology and the practical issues that it generates.

The more immediately apparent environmental relations and impacts of wood, that we previously described as a 'transparent ecology' (see Chapter 3) can be a powerful pedagogical tool for giving students a more critical understanding of the meaning of sustainability in architecture.

5.1 BEYOND NORMATIVE WAYS OF TEACHING ARCHITECTURE

As a world-weary Professor of Architecture once confided, perhaps the true intention of architectural education is not to teach students how to make buildings, but how to make drawings of buildings. A student of architecture can pass their entire education without touching any of the materials used in the making of buildings or visiting a construction site. Since the emergence of the modern architect in Renaissance Italy, the role has progressively detached from the messy business of making buildings. As a result, professional architectural education has evolved to reinforce the idea that an architect designs and others build. While the content of architectural education is often determined by professional accrediting bodies, the possibility of working hands-on at a one-to-one scale with building materials remains a diversion from normal curricula.

The education of an architect is also a lengthy and demanding process, one that is far from complete on graduation day. Students typically spend five years in full-time education before beginning a period of supervised practice or apprenticeship. In the United Kingdom, an architect is only qualified after several years of practical experience that has been accounted for, reflected on, and tested in the Part III examination administered by the Architects Registration Board (ARB)/Royal Institute of British Architects (RIBA). Such is the expectation of a long process of continued learning, that the magazine *Architects Journal* has for many years organised the 40 Under 40 award for 'emerging' architects.[1] Architecture may be a discipline that takes a long time to master, but at least you are still considered young until your fifties.

On this long journey, where does a student learn about the characteristics, capabilities, and potentialities of wood as a material? Although architectural education curricula vary from country to country and university to university, there are patterns that are common to the curricula in most schools which we will now consider as the normative (i.e., standard) model of architectural education.

In normative architectural education, approximately half of a course's credits and study time will be spent in some form of design studio course, in which students synthesise their learning through iterative design projects. Students will be given a brief of varying detail or complexity, and they will respond to that brief (over a period of time as short as a day or as long as an entire semester) with a design proposal. In tutorials, desk crits, or more formal critiques, tutors might suggest examples of buildings for the student to study as a design precedent in pursuit of their own design project. The design project becomes a means of synthesising the many aspects of architectural practice. This synthesis is undertaken through the representation of drawings and scale models, either physical (see Figure 5.1) or digital.

In the remaining fifty per cent of their studies, students might see historical precedents of wood buildings in lectures – for example, learning about how the classical language of Greek temples built in stone is derived from their long-lost wooden forebears. In structure or technology courses, students might learn about the characteristics of standardised wood products and the effects of bending moments on timber elements. In the workshop, if they have access to one, students might get to handle smaller pieces of wood to construct models or representations of their designs. Small wood elements can be used to create frames, grids, or matrices to test structural strength.

The educational initiatives illustrated by the case studies in this book, all of which give students of architecture the chance to practise working hands-on with wood, share a criticism of the normative architecture pedagogies described above. These case studies represent the efforts of individuals within architectural education to improve, enhance, and develop new pedagogies which give students a better understanding of the qualities of wood as a building material as part of their education. As Piers

Figure 5.1 Students at Umeå School of Architecture review cardboard models produced over the course of one day in a design course. Photo: James Benedict Brown.

Practising Wood in Architecture

Taylor, co-founder of the Studio in the Woods explains, working hands-on with a building material is:

> … incredibly different from using something that's an abstract idea in studio … When it's tangible and real, I think it becomes a generator for the project rather than something that you just apply afterwards when you think about how you might make your abstract idea.

The next parts of this chapter explore how the different case studies in this book interpret the possibilities and limitations of teaching with wood and some of the specific pedagogical innovations demonstrated by our examples.

5.2 NEW ROLES FOR WOOD IN ARCHITECTURAL EDUCATION

Just like any other material, wood has no pre-determined function (Ingold, 2021). How we see it, use it, and apply it reflects not only its physical properties but also what we see in it and what we ask of it. If we approach wood as an industrially produced material available in standardised shapes, sizes, and strengths, then we are unlikely to use it any differently. Likewise, if we approach wood as an answer to an environmental crisis, then we will inevitably see the material through that lens.

Through the case studies described in this book, we see that wood has the possibility of becoming something else: an active and reciprocal part of a student's educational journey, rather than an inert material to be selected from a catalogue. In many of the examples in this book, we can see teachers and students creating a dialogue with the material, engaging with the problems and processes of design and construction in a way that is more sensitive to the possibilities of the material. Furthermore, there is a very different relationship between the designer and the forest in the university than in the industry. In the university, material choices and design decisions are made according to a set of parameters that prioritise learning, research, and experimentation. Ethical considerations can also be prioritised over cost, profit, and loss, which play a greater role in commercial settings.

The role of wood can be more explicitly understood in the instances where a deliberate choice has been made to start using or switch to using wood instead of other materials. The shift to timber observed in the University of Nottingham's (UoN) Design+Build Studio is justified in terms of wood's accessible, democratic, efficient, and forgiving characteristics. Timber-building components are generally easy for students to move and lift safely. Compared to buildings made with other materials, wood buildings are easier to build, modify, and take down.

That ease of use is also evident in the work of the Insitu Project. Its participants have developed a similar approach to wood: using it as a simple, cheap, quotidian material that is easily acquired on site in China, but which also has a rich tradition of Chinese carpentry and woodworking skills behind it, some of which are at risk of being lost if they are not practised.

During the long weekend spent at the Studio in the Woods (the educational initiative with the shortest duration of all our case studies), the speed, accessibility, and versatility of wood, which requires only simple tools and can be handled by someone with almost any skill, is key. The rapidity with which groups at the annual Studio in the Woods workshop can produce ambitious, elegant, and sophisticated structures in just a couple of days illustrates the ease-of-use that is characteristic of wood. Several of the respondents to our research have described these characteristics as integral to wood's openness, directness, and even its disregard for skill or experience.

At both Studio in the Woods and the Architectural Association's (AA) Hooke Park Design+Make programme, two projects that share similar environmental settings but have different durations, architecture students are given the opportunity to not only work with but also live in the forest. They can work with recognisable wood products but also with raw materials, small-diameter wood, unseasoned wood, and materials normally discarded in industrial forestry. One of the goals of the programme at Hooke Park is to avoid the unnecessary felling of trees and to get as much use as possible out of the materials that are harvested. Such platforms give students a better opportunity to understand the origins, species, and characteristics of trees and the wood products they produce. In the words of Piers Taylor, recalling his first experience of felling and milling timber from his own land:

It was mind blowing that you take a tree that architects historically would just cut into regularized pieces and throw two-fifths of it away, to see the waste, see the actual material. You know, a tree, of course, is never straight. It's always kind of wonky. And it may sound obvious just to say it, but suddenly being aware that our own presumptions around material weren't abstract things but rather real things. We had a scarce amount of material and we had to understand how to use it intelligently.

These projects critique and even reverse an extractive approach to the forest, prompting critical engagement with the methods of producing wood for construction and aiming to nurture a relationship with wood that is more dynamic, active, and engaged. Such an approach is necessarily critical of more wasteful production processes and the standardised, industrially produced timber we are accustomed to using. What this requires is a change in attitude: what one respondent described as asking 'what the forest offers versus what I want from the forest.'

Some of our case studies, such as the Design+Make Masters courses at Hooke Park and the Wood Program at Aalto University, are located within the second-cycle of architectural education, which has a greater expectation of engagement with research from students. In these examples, it is not the simplicity of wood but its complexity and potential as an industrial product that drives innovation. The advanced computational and manufacturing processes demonstrated by the work at the Institute for Computational Design/Institute of Building Structures and Structural Design (ICD/ITKE) in Stuttgart show how second-cycle students can engage with more complex wood prototyping and manufacturing products. Gramazio Kohler's research makes clear how timber can allow for complex geometric design and robotic fabrication. Different woods function differently in these processes, so there are further opportunities to study and understand the performativity of various kinds of wood.

5.3 HANDS-ON LEARNING

Whereas with materials such as masonry or steel, where you have a limited range of processes, mostly subtractive, that you can use on them, or with concrete where you have a very plastic material, but you need to put the majority of your effort into the container that will shape it, into the formwork. With timber we can design directly for fabrication, be they additive processes like laying down additional timber members or timber beams and connecting them, or subtractive processes like milling.

In the extended quote above, Luis Orozco concisely explains why, for pedagogical reasons, wood has certain characteristic advantages over other materials. Wood is ideally suited to hands-on learning of architectural design and construction precisely because of the opportunity to 'design directly for fabrication.' The educational initiatives described in this book connect the processes of designing, researching, and making, giving students the opportunity to work hands-on with the materials and processes of building and making architecture a tangible process.

Different case studies in this book demonstrate how students of architecture can profit from a critical hands-on engagement with standardized wood products. For UoN's Design+Build Studio, using only standardised timber products facilitates construction and increases the likelihood of completion on a tight budget and timescale. That just two undergraduate students of typical strength and ability can lift a piece of timber above their heads during the roof construction phase of the studio's projects in South Africa appears critical to the continued success of the initiative.

In our interviews with them, both Peter Russell of the UoN's Design+Build Studio and Luis Orozco of ICD/ITKE expressed similar sentiments about the importance of connecting the act of designing a building with the act of handling the components. Orozco explains that 'one of the main educational goals… is to understand that what you draw has to be buildable and that if you are thinking about how it is produced as you are drawing it, you are making a better architecture.'

Similarly, for Emmanuel Vercruysse at Hooke Park, 'any aspects or any initiatives that narrow that gap between the actual physical production of architectural components and architectural education can only be beneficial for [students'] education' (emphasis added). The gap which Vercruysse describes is the distance between architectural education and practice which has existed since the discipline first professionalised and then entered the university. As the role of the architect moved away from the master-builder and as the training of the architect moved away from apprenticeship to university education, a distance opened up between theoretical knowledge and hands-on knowledge of designing buildings. In these educational initiatives, we see efforts to correct this: design-build projects are attempts to repair a broken or loose connection between design and construction.

Learning hands-on with wood is not necessarily a natural process. Vercruysse describes the 'unlearning' that second-cycle (graduate) students with three or five years of normative architectural education have to go through. Students can be hesitant at first:

> ...because this is not necessarily their comfort zone or a natural environment to explore things through making. But I think that what is really fascinating is to see the evolution of the students from not necessarily being confident in workshop environments to actually see that shift.

The educational initiatives described in this book also suggest a critique of broader educational trends, notably the lack of opportunities for primary and secondary school students to handle materials and experience making before they go to university. Peter Russell, who has been involved in design-build teaching in both the USA and UK, reflects on how:

> ...in the 1980s, [shop class and woodworking in US secondary schools] were all cut based on budgets and health and safety, and so just now we're starting to get to this generation of architects who are in charge of big projects and have never built anything.

While some students will continue to benefit from access to well-equipped workshops and high-quality teaching before they come to university, working hands-on with wood through the initiatives described in this book can give students an opportunity to correct inequalities of opportunity created in secondary education.

For many educators, the ambition of many of the projects described in this book will feel incompatible with the workshop resources available to them. It is worth noting that across all of the case studies presented here, there is a wide variety of workshop resources. While the AA's Hooke Park campus and Gramazio Kohler Research (GKR) can use large and well-equipped indoor workshops for testing and pre-fabrication, smaller initiatives like the Studio in the Woods usually operate with only hand tools on site. The UoN's Design+Build Studio is able to prototype certain details and structures on campus, but the sum total of its equipment has to fit inside a shipping container that is moved from building site to building site each year. On this varied access to workshops, Peter Hasdell, of the Insitu Project, speaks fondly of the widespread provision of high-quality workshops in North American and European schools of architecture, notably (irrespective of space or size) the 'much more active workshops where students would be able to cut and make and join things together themselves, and they were supervised, but there they were encouraged to do those things.' In this reflection, we are reminded that a culture of making and experimenting in the workshop is characterised by the level of activity and engagement, not necessarily by the availability or cost of tools and equipment.

5.4 ECOLOGICAL AWARENESS

In the opening chapters of this book, we remarked that there is no greater challenge facing architecture students today than resolving the excessive environmental impact of our profession. We have also argued that wood shows a transparent ecology, underlining how the material and temporal scale of its life cycle are more easily graspable when compared to other building materials. In this part of the chapter, we want to continue to show that the hands-on experience of working with wood (i.e., in design-build projects or hands-on experiments) can be a legitimate and productive means of connecting the individual and the collective awareness of our ecological responsibilities.

Since the 1992 Earth Summit in Rio de Janeiro, sustainability has most often been defined according to three equally weighted dimensions: economic, social, and biological. This can lead to tensions between public, private, and governmental actors, because it positions economic sustainability on equal footing with environmental sustainability. Stakeholders in forest industries have argued, therefore, that industrial methods of forestry can demonstrate economic and social sustainability which trump the biological concerns of clear-cutting and monocultural replanting. While many of our respondents acknowledged this tripartite understanding of sustainability, they also expressed very particular opinions about the ecological awareness that comes from working hands-on with wood. It should also be stated clearly that wood is not sustainable by default: sustainability depends on a critical understanding of where the material being used has come from and where it will go at the end of the building's life.

Much of the motivation to increase the use of wood in construction comes from wood's ability to sequester atmospheric carbon. The forest industries are naturally interested in increasing consumption of their products, but some of the initiatives studied in this book demonstrate a different approach, one which is mindful of the impact of increased forest production. Anna Krtschil clarifies that at ICD/ITKE 'our goal is still to use as little material as possible and not as much as we can, because this would of course increase the carbon storage capacities of the building but [it would] also impact on the forests.'

Reflecting on recent trends for high-profile and high-budget wood buildings, ones which use large volumes of wood and demonstrate significant capacities to sequester atmospheric carbon over the lifetime of the building, one respondent reflected that, nonetheless, 'more money doesn't buy you better two-by-fours.' While engineered timber expands the possibilities of building with wood, there remains a certain democracy to wood's inherent sustainability. Similarly, a respondent noted that 'we exist in [an institution] that is obsessed with solar panels and insulation … we are the outliers saying, you know, there are 17 UN sustainable development goals and energy is only one of them.' For architecture educators working with wood, the material is a threshold to broader discussions about ecological awareness and sustainability.

Anna Krtschil's colleague at ICD/ITKE, Luis Orozco, expanded on this observation by noting that a guiding strategy for sustainability is to reduce the volume and number of metal components as much as possible. On the way to realising this ambition, the BUGA Pavilion is an example of a compressive structure built with steel bolts but which the designers are confident could stand without them. At the time of our interview, Orozco cited work-in-progress that would provide 40 square meters of construction using less than six linear meters of one-centimetre diameter steel. Aiming to eliminate the use of steel or adhesives in construction remains a compelling motivation for wood projects, especially smaller pavilions built by students that may have a second life when they are disassembled or moved elsewhere. An ecological awareness of the potential sustainability of wood must, therefore, take into consideration the fixings and adhesives chosen. Reflecting on the progression towards more ecologically sensitive adhesives in manufactured timber products, Yves Weinand of IBOIS (the Laboratory for Timber Constructions at the École Polytechnique Fédérale de Lausanne) reflects that:

> …lots of timber products in the 1970s and 1980s were not so sustainable, you know, with the glue in it … we still have the glue, which is used in LVL (Laminated Veneer Lumber) panels. That glue has improved over time, but we still have that.

In this context, Weinand explains that buildings should be environmentally costed according to not one, but three cost estimates: the environmental cost of the construction and operation of the building itself, the cost of the embodied or grey energy consumed in the production of materials, and the environmental cost of the dismantling, demolition, or recycling of the building. Wood has the potential to have a lower ecological impact than steel, concrete, or masonry, but only if these different ecological costs are taken into account.

Speaking from the perspective of a campus established more than four decades ago, Emmanuel Vercruysse reflects that:

> …the assembly logic is really half of the design project [and] … the afterlife of these materials is really essential …the assembly logics, mechanical fixings, other glues and everything is analysed and scrutinized to make sure that they don't become a problem when you look down the chain.

The integrated approach to forestry, design, and construction being prototyped at Hooke Park is based on a rejection of the notion that some forest materials have zero value. A more ecologically aware approach to designing buildings is to recognise that all materials, even those typically discarded before they are offered to the architect, have a value of some kind (and not, as is widespread in Swedish forestry, as fuel for energy production).

The strong connection that Hooke Park has established with the forest on its own land is also an example of how – even in the limited duration of a single academic year – the transparent ecology of wood can be an effective pedagogical tool, making students aware of the full life cycle of the material and, just perhaps, its ecological complexities. This awareness is not necessarily limited to wood: if understood holistically, it could then be applied to other building materials and processes, giving students the critical tools to foster sustainability in all aspects of their practice.

5.5 WOOD AS A MULTIDISCIPLINARY MATERIAL

In the research informing this book, we found many instances of advocates of working hands-on with wood in architectural education who referred to their work as multidisciplinary, interdisciplinary, or even transdisciplinary. There are at least five common English prefixes to the word 'disciplinary' which are used to describe initiatives that involve participants from different disciplines. It is useful to disambiguate them in order to highlight some of the characteristics of the initiatives surveyed in this book.

Figure 5.2 is a diagram drawn after that of Professor Alexander Refsum Jensenius, which in turn was drawn after that of Marilyn

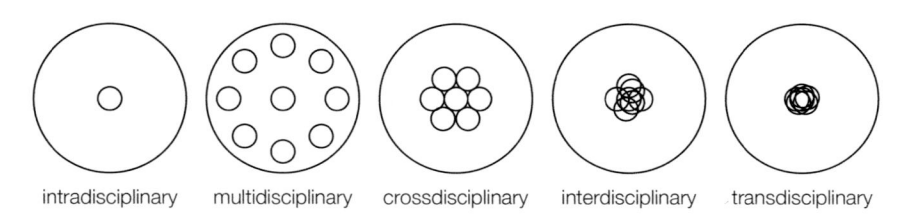

intradisciplinary multidisciplinary crossdisciplinary interdisciplinary transdisciplinary

Figure 5.2 Visualising intra-, multi-, cross-, inter-, and transdisciplinary structures. Diagram: James Benedict Brown and Francesco Camilli, after Alexander Refsum Jensenius.

Stember (1991). Jensenius distinguishes five levels of disciplinarity.

- Intradisciplinary: working within a single discipline.
- Multidisciplinary: people from different disciplines working together, each drawing on their disciplinary knowledge.
- Crossdisciplinary: viewing one discipline from the perspective of another.
- Interdisciplinary: integrating knowledge and methods from different disciplines, using a real synthesis of approaches.
- Transdisciplinary: creating a unity of intellectual frameworks beyond the disciplinary perspectives.

Reminiscent of Sherry R. Arnstein's *Ladder of Citizen Participation* (Arnstein, 1969), we see an implied positive progression from intradisciplinary to transdisciplinary practices. The implication of transdisciplinary practices is a move towards the complete integration of different disciplinary structures, perhaps even to a new unified disciplinary framework (which might return us to the beginning of the diagram).

Initiatives in architectural education that engage people from different disciplines have the potential to reach all five levels of disciplinarity described in Figure 5.2. However, to learn from the widest possible range of precedents, and to avoid prejudicing those initiatives which have had less time to develop their practice, we have conservatively focused on any initiatives that show evidence of at least multidisciplinarity. A multidisciplinary project is one in which people from different disciplines work together – during this collaboration each draws on their own disciplinary knowledge in pursuit of a common outcome.

In our diagram, we can imagine each of the five bounding circles to be a project. Within each circle are individuals and groups with different disciplinary expertise. They work together, in proximity, and they deliver an outcome that represents the sum of their collaboration. That is not to say that there is no potential – nor evidence in some of our case studies – for teaching and research projects that involve truly interdisciplinary or transdisciplinary work. Such examples can be found in the second-cycle projects that engage students in original research that can generate new knowledge beyond the frameworks of a single discipline. But for the purposes of our study, understanding the opportunities for hands-on experience with wood in architecture education, we are most interested in multidisciplinary projects that demonstrate to architecture students the contribution of the different disciplinary knowledge frameworks. Students and teachers of architecture enter the initiative as architects and emerge from the other end as architects, still, but hopefully with a greater understanding of the contributions and expertise of the other participants.

At Aalto University in Finland, architecture, civil engineering, and structural engineering students collaborate in the Wood program. This is deliberate rather than tokenistic, engaging with faculty members from allied departments as early as possible in the process. Yves Weinand of IBOIS in Switzerland directly acknowledges the inspiration of Pekka Heikkinen, a founder member of the Aalto Wood Program. The formal multidisciplinary collaboration of the Wood Program was pioneered in a more modest summer workshop uniting architectural and technical approaches to working with wood: 'he went a little bit further because he also really tried to tackle constructive and technological aspects within the summer workshop, much more than it was the case for others.'

Today, IBOIS prides itself on its multidisciplinary collaborations that are undertaken with actors in different parts of the construction industry. IBOIS is both an educational initiative and an architectural practice. Working with the industry suggests not only that the role of the architect is becoming more collaborative, but also that the role of the architect might need to evolve, to become involved differently in the design, fabrication, and construction processes. Weinand refers specifically to a successful collaboration with a local timber company, which through their work with IBOIS has now expanded into CNC production.

Pleasingly, we see in the case studies evidence of how teachers can bring about true multidisciplinary collaboration by starting small. The different scales of the case studies in this book, from a short weekend workshop to second-cycle (graduate) programmes, should not be seen as static: many of the initiatives described in this book started small, and they continue to inspire for that reason.

For the reader interested in establishing their own multidisciplinary projects, perhaps as a teacher of architecture, the tendency of higher education faculties, departments, and programmes to be siloed from one another can be disheartening. But the reader will find in the different case studies in this book varying examples of multidisciplinary collaboration that can provide templates for their own initiatives.

Note

1   See https://www.architectsjournal.co.uk/news/40-under-40.

Bibliography

Anderson, Jane. 2019. 'Live project/designbuild education: Creating new connections between universities, communities and contemporary professionalism.' In *Defining Contemporary Professionalism: For Architects in Practice and Education*, edited by Alan Jones and Rob Hyde, 5–8. London: RIBA Publishing. https://doi.org/10.4324/9780429347856-2

Arnstein, Sherry R. 1969. 'A ladder of citizen participation.' *Journal of the American Institute of Planners*, 35(4): 216–224.

Ingold, Tim. 2021. *The Perception of the Environment: Essays On Livelihood, Dwelling and Skill*. London: Routledge.

Case Study 5

Wood Program in Architecture and Construction, Aalto University, Finland

DOI: 10.4324/9781003428930-10

Figure CS5.1 Scale model of the Kide pavilion, Kuhmo, eastern Finland, 2024. Photo: Darya Belaia.

Finland is the most forested country in Europe, with three-quarters of the land mass of the sparsely populated Nordic country covered in woodland. Growing and developing timber construction in Finland is a stated mission of the Finnish government, and the Ministry of the Environment's Wood Building Programme (2016–2023) has sought to increase the use of wood in urban development and in public buildings as large constructions (such as bridges, sports halls, etc).[1] Given this context, it is unsurprising that Finland is home to some of the world's most cutting-edge educational programmes and research in the field of wood architecture.

Named after the renowned Finnish architect Alvar Aalto (1898–1976), Aalto University is a relatively young institution, formed in 2010 as a merger of three separate higher education establishments: the Helsinki University of Technology, the Helsinki School of Economics, and the University of Art and Design Helsinki. The School of Arts, Design, and Architecture is one of six schools in the university (and one of only three architecture programmes in the country).

The one-year, graduate-level (60 ECTS) Wood Program in Architecture and Construction was established prior to the merger in 2000 as an initiative of the former Department of Architecture. The programme unifies previously unconnected courses on the technology, history, and design of wood. At the time of writing, the programme is overseen by Pekka Heikkinen, who is Aalto University's first Professor of Wood Architecture. A multidisciplinary profile is maintained with the formal engagement of teachers and students from the Department of Engineering, and specifically students of civil engineering.

While Aalto's Bachelor and Master's programmes in architecture both focus on the materiality of architecture, the one-year Wood Program is focused exclusively on understanding wood from a variety of critical perspectives, including forest management, ecology, and sustainability. Good connections with industry support the programme, and at the time of our research about five species of wood grown in Finland are being used in the programme. Some imported material is occasionally used, but it is a point of pride to use only Finnish material and to explore its fullest potential through design and research. In the first assignment, students are asked to work hands-on with several different species in order to get a better understanding of their different properties. As the programme moves towards the design, prototyping, fabrication, and construction of a structure, students work mainly with Finnish spruce, pine, and industrial products like plywood or glulam. Built outcomes vary from year to year depending on the brief. In some years small pavilions are constructed for specific events, whereas in others students might be tasked with designing a small unit for volume production.

The course provides both basic and technologically advanced tools to work with, so that students can be exposed to the variety of possible uses of wood. Occasionally, some more complex manufacturing can be outsourced to industries with the appropriate capacity.

The properties of timber guide the design process and the spatialities explored in the projects. This is coupled with a practical research topic that is changed every year, giving students a common focus. Students are stimulated to find references for their design proposals in vernacular architecture and update them to work with modern technologies. They are also invited to use wood as much as possible – for example, by experimenting with wood–wood connections. Through iterative tests, they explore design alternatives that can increasingly optimise material use.

By exploring the properties of timber, students are expected to become practitioners with a critical consciousness of the implications of the use of timber in a building. This is achieved by exposing students to the material in a structured way: in the initial phase of the course, they are challenged with simple tasks, like, for example, working with a small wooden cube to familiarise themselves with the material, exploring ways to cut it or performing simple compression and tension experiments to observe how it cracks. Practical experiments are then complemented by lab tests and mock-ups.

Wood integrates into the pedagogic methods of the programme so well because it allows students to work additively by

Figure CS5.2 Kide pavilion under construction, Kuhmo, eastern Finland, 2024. Photo: Daniel Sars.

Case Study 5: Wood Program

Figure CS5.3 Kide pavilion inauguration, 2024. Photo: Meina Kobayachi.

assembling small elements. In addition, its hygroscopicity makes it more informative for students who can observe its reactions to changes in the surrounding environment.

Sustainability is incorporated by looking at the entire lifecycle of the material, from harvesting to disposal. This implies a particular focus on detailing the project, to allow for easier disassembly but also for sustainability in the use phase.

The buildings realised by the Aalto Wood Program are usually publicly accessible community projects, mostly with municipalities as clients. In 2024, the Wood Program celebrated its thirtieth year with a design-build project for a public square in Kuhmo in eastern Finland. Such projects go through participatory processes and dialogues with future users, promoting their integration into the social and urban context. The programme also seeks collaboration with the industry, integrating their research skills with productive capacity.

Note

1 See https://ym.fi/en/wood-building.

Case Study 6

Design+Build Studio, University of Nottingham, England

DOI: 10.4324/9781003428930-11

In the 2008/09 academic year, a large multidisciplinary project drawn from across the Department of Architecture and Built Environment at the University of Nottingham (UoN) in England designed and built a crèche in the community of Jouberton in Klerksdorp, North West Province, South Africa. The first project in Jouberton was widely published and recognised for its innovation, although as an academic endeavour it was beyond the capacity of the department to repeat it sustainably every year.

After a brief pause, the Design+Build Studio returned to South Africa in 2011, completing its first project in Limpopo. By the university's own estimates, the studio has engaged more than 300 students of architecture in the design and construction of accredited childcare facilities in South Africa. The studio has built these crèches with the collaboration of two non-governmental organisations (NGO): Education Africa, a non-profit founded in 1992 focused on poverty alleviation through education, and, since 2014, the Thušanang Trust, founded in 1986 and established as charitable trust in 1992, in Haenertsburg, Limpopo Province. The Thušanang Trust supports the well-being of young children through the development of accredited Early Childhood Development (ECD) centres and the training of teachers.

Figure CS6.1 Monang Preschool, 2023. Photo: Thabang Radebe.

Many tenured and visiting staff of the UoN have contributed to these projects, including consultant structural engineers who give their time *pro bono*. We interviewed the studio's current director, Peter Russell, who joined the department as an assistant professor in 2017 and assumed responsibility for the Design+-Build Studio shortly thereafter.

Russell made some changes to the programme, including a concerted decision to build all future projects with wood. Russell acknowledges that this shift in direction was influenced by his background in construction and design-build pedagogies in the USA, where timber is a much more commonly used material in domestic scale buildings than in the UK or South Africa. But he also argues that timber has certain qualities that make it very democratic:

> We can skill up basic timber skills to build a stud wall or a structural wall out of timber, especially if you've got an engineer or a lecturer or a builder guiding you. We can teach young people who've never done anything [like this before] how to put nails together correctly very quickly. We can

Figure CS6.2 Monang Preschool, 2023. Photo: Thabang Radebe.

teach them how to do sheeting very quickly, and they can lift and move this stuff very quickly. It's very efficient in terms of its material properties for working with young people. It's very forgiving and, at the end of the day, you get from building material to a pretty high-spec finish with timber. The learning curve is shorter than with a lot of other materials.

With more than a decade of experience delivering projects on the other side of the world, the UoN Design+Build Studio has no precedent in British architectural education.

In our interview, Russell explained (with our emphasis) that students taking the Design+Build Studio elective in their second year of studies must understand 'not just *what* we build but how we build it and *why* we build.' Russell argued that students are sensitive to work that is meaningful, explaining 'they know when they're being given a piece of project work that their tutor has invented and is going to assess based on whatever he or she likes [it].' A design-build project that provides a new childcare facility for a community in Limpopo 'is going to make the lives of a lot of people measurably better … The students can spot it … there's something about the projects that makes it a little easier to sleep at night if they're done well.'

The studio allows students to engage hands-on with issues relating to the United Nation's Sustainable Development Goals: equality, access to education, access to water, and community empowerment. But in our interview, Russell acknowledged that the Design+Build Studio 'is not perfect by any means,' and that continuing effort was needed to ensure a decolonisation of the curriculum.

Russell emphasises that the success of the Design+Build Studio's projects is dependent on a recognition of the importance of the task at hand: it must be finished by the handover date, and failure is not just unacceptable, it is unethical. A few deliberate steps have been taken to increase buildability and to reduce the risk of a failure to deliver a completed building. Groundworks, for example, are now subcontracted to local contractors. Since 2018, students have arrived on site to find cured concrete pad foundations ready for their timber construction, saving time, ensuring a better-quality foundation, and diversifying the financial impact of

Figure CS6.3 Groundworks, including concrete foundations, are now subcontracted before students arrive on site. Monang Preschool, 2023. Photo: Thabang Radebe.

Figure CS6.4 Monang Preschool, 2023. Photo: Thabang Radebe.

the project amongst local businesses. Another recent development has been the introduction of a handover celebration on the last day of construction. The celebration is important for students, teachers, and the client to mark the end of construction. But it also takes on a new significance: at this moment, the leader of the crèche takes ownership of the building. It has been observed that without a clear moment of handover, the commitment of the university and its responsibility towards earlier projects (many of which now require preventative or corrective maintenance) is unclear.

Chapter 6

The Ecology of Wood

Wood's potential to be a sustainable building material is conditional on the preservation of the complex web of ecological relationships of which it is part. At the 1992 United Nations Conference on Environment and Development (UNCED) – or Earth Summit – in Rio de Janeiro, member states of the United Nations (UN) agreed on the Rio Declaration on Environment and Development and the UN's Agenda 21. The conference also provided a new definition of sustainability, written in Principle 3 of the Rio Declaration: 'the right to development must be fulfilled so as to equitably meet developmental and environmental needs of present and future generations.' The UN's definition of sustainability set out to recognise the interconnectedness of economic, social, and environmental aspects of human well-being, as well as the importance of balanced but continued development. As noted in Chapter 5, by considering the needs of current and future generations, the UN's definition of sustainability introduced a tension between public, private, and governmental stakeholders, defining economic sustainability on an equal footing with biological and social sustainability.

This definition emphasises the complexity of the concept of sustainability, and the many interconnected aspects that need to be considered in order to achieve it. This web of relations can be understood as an ecology.

DOI: 10.4324/9781003428930-12

Figure 6.1 A participant at the UN Earth Summit (3rd of June 1992) signs his name to the Earth Pledge in which he promises to act to the best of his ability to 'help make the Earth a secure and hospitable home for present and future generations.' Photo: UN Photo and Michos Tzovaras.

According to Encyclopedia Britannica:

> …the word ecology was coined by the German zoologist Ernst Haeckel, who applied the term *oekologie* to the 'relation of the animal both to its organic as well as its inorganic environment.' The word comes from the Greek oikos, meaning 'household,' 'home,' or 'place to live.' Thus, ecology deals with the organism and its environment.
>
> (Smith and Pimm, 2024)

In Chapter 3 we explained how the ecology of wood is easier to grasp when compared to the ecologies connected to other non-organic materials. In this chapter, we further explore the relationship that wood as a material has with its context, encompassing not only environmental aspects but also social and cultural ones. We underline how wood interacts with changing contextual conditions, how it connects to local cultures and

Practising Wood in Architecture

knowledge, and how its use can affect local communities. We also look into how its material behaviour and the implications of its use can emerge more clearly from more conscious research and practice.

6.1 MATERIAL ECOLOGIES

Wood can be employed in construction in a variety of different ways. One can imagine a spectrum ranging from the simplest application of a raw material to the most complex application of an engineered material, from a log cabin to an engineered mass timber building with spans reaching further than a normal beam thanks to engineered timber products. One is not necessarily better than the other: these are just different ways of using wood that reflect different economies of production. In each instance there is a different interaction with the material. The log cabin might use more material, but it requires less manufacturing and processing. Wood lends itself to a variety of different applications, some requiring more or less material and some requiring more or less processing (from raw material to finished object).

Figure 6.2 Designing with the irregular shapes of nodes between tree trunks and branches. Photo: Postgraduate Design + Make course, Architectural Association.

The availability of material, time, and money play a part in these decisions.

As an organic and heterogeneous material, wood is subject to a variety of grading mechanisms to meet the standards of industrial production. Irregular elements, like the nodes between branches and tree trunks, tend to be discarded. Consequently, the proportion of biomass used for construction timber can be as low as 30 per cent. The wood that becomes construction timber can also be subjected to additional applications or impregnations of epoxies, glues, and other chemicals to manage wood's natural hygroscopicity. These chemicals and non-wooden elements, such as steel fastenings, all make it harder to recycle wood at the end of a building's life cycle. Introducing students to methods of eliminating these additives can not only simplify the building process and reduce waste and pollution, but also leads to remarkable creativity.

When students of architecture can not only work hands-on with wood but also in the forests that produce that wood, opportunities present themselves to expand their understanding of the material economies of wood. At the Architectural Association's (AA) woodland campus, Hooke Park, the buildings of the campus itself demonstrate ways of using building materials usually rejected by normative production chains. By surveying the woodland around the campus, both by eye and using digital tools, students can witness new ways of identifying irregularly shaped trees and younger saplings that have characteristics that might be used in each application. While the possibility of owning established and mixed forest land feels like a privilege few architecture schools can entertain, the beauty of much of the work produced at Hooke Park lies in its material economy. Small, twisted, or bent trees would be discarded or thinned out of conventional forestry production flows, potentially sent for use as pulp or biomass material or left to break down on the forest floor. These are, by normal measures, cheap or even worthless pieces of wood. Hooke Park's approach to the forest expands the material economy of wood and challenges a lazy industrial dependency on the standardisation and homogenisation of wood.

There are two standout examples from the Hooke Park campus. The Woodland Cabin (designed and built by students Wyatt

Armstrong, Ciro Romer, Hilla Gordon, Nasia Pantelidou, and Shengning Zhang in 2020) is a small, intimate, and atmospheric space for educational activities that looks out onto the forest landscape. Closer to the centre of the campus, the Wood Chip Barn (designed and built by students Zachary Mollica, Swetha Vegesana, Sahil Shah, Vivian Yang, and Mohaimeen Islam in 2016) is a much larger building that employs sections of 20 forked beech trees from the woodland around the campus. Working in collaboration with engineers from Arup, three-dimensional scans of 25 trees were processed by a digital script to develop a structural concept. A digital model and robotic arm were used to machine 20 of these into the finished building components. Today, the building not only serves as a demonstration project for the productive use of irregular wood components and digital modelling technology, but also provides more than 400 cubic metres of storage for wood chips that are produced on site and used for heat production. While this project made extensive use of cutting-edge technology to scan and manufacture the individual building elements, the principle of studying the forest to identify and design with irregularly shaped sections of trees should not be considered out of reach to those without access to the same technology. Similarly to the AA's Hooke Park campus, at the (itinerant, temporary) Studio in the Woods a programmatic choice is made to source materials from the forest in which work is undertaken. The entire initiative is shaped by a creative engagement with a material constraint. Having a limited choice and quantity of raw material challenges students to find innovative solutions while also considering the ecology of the process of building.

The application of technology during the design phase can enable students to maximise the productive value of the raw material they use. Our study of the joint projects delivered by the Institute for Computational Design (ICD) and the Institute of Building Structures and Structural Design (ITKE) in Stuttgart highlights ways in which to expand the optimisation of wood in lightweight shell structures and multistorey timber buildings. Multistorey and tall timber buildings are increasingly being advocated for on the grounds of their potential to capture and store large volumes of carbon. But using more timber just for the sake of increasing carbon capture capacity is a contradictory approach. Incentivising the use of greater quantities of timber places additional pressure on forests and ecosystems. In our research, we found respondents from ICD/ITKE were eager to articulate that they were pursuing optimisation of material use through the effective exploitation of its physical properties. These projects explore optimisation of both material and space, leading to lightweight structures adapted to denser and therefore more efficient inhabitation.

Some of the case studies in this book engage with a greater degree of social responsibility, designing and realising buildings that represent a *pro bono* contribution from the university to the community that receives them (we look in more detail at the social ecologies of wood later in this chapter). In these instances, the need to achieve high-quality results with limited or strictly constrained resources – material, time, and skill – is of great importance. Whereas we have seen how students at the AA's Hooke Park campus can use cutting-edge technology to scan and digitally model structural systems using irregularly shaped pieces of wood, there is a no-less important role in student design-build projects for standardised pieces of manufactured wood. In the suburbs of Ulaanbaatar, Mongolia, the teachers and students of the Rural Urban Framework (RUF), in collaboration with GerHub, delivered the Ger Innovation Hub (GIH) to contribute to improved social conditions for internal migrants. Using timber to construct a lightweight frame, it was possible to build a double-layered structure with insulated spaces, designed for a variety of activities, that remained comfortable throughout the frigid Mongolian winter. In South Africa, the University of Nottingham's (UoN) Design+Build studio has delivered timber-framed childcare centres that can significantly change the quality of life of local communities. These trans-national projects are realised with limited funding, usually raised by the students themselves, and with limitations of time and skill that come from working with undergraduate architecture students. In this context, standardised timber components can enable a materially efficient building, as they can be easily designed for by students working in separate off- and on-site phases, and they can usually be sourced from multiple retailers at competitive prices.

As discussed in Chapter 5, wood is not ecologically sustainable by default. The ways trees are grown and harvested and the ways in which wood is manufactured and used in buildings all contribute to its sustainability. We have seen, in the preceding examples, a variety of different ways in which students can work hands-on with wood and engage with the material economy of wood. In very different ways, these projects help students to become more aware of where wood comes from (or how it might be procured as a standardised and globalised industrial product) and how that availability might be used efficiently.

6.2 LOCAL ECOLOGIES

As discussed in Chapter 4, wood as a material has the potential to have deep and meaningful connections with its local context. The use of local materials with their own aesthetic characteristics can support a more harmonic integration of buildings into the built and natural environment (especially if existing buildings are constructed in the same materials). Forests define the landscapes in which they sit, and the many dimensions of forestry and forest activities contribute to the formation and preservation of cultural identity. The case studies presented in this book vary in the intensity of their relationship to their local environment, which offers the reader provocations regarding how we conceive of the local ecology of wood. Many projects with this degree of contextual engagement are working not with new-build construction but with repair, restoration, or adaptive reuse of existing buildings, which is an important aspect of architectural practice today that is often side-lined in normative architectural education. In some examples, the work of teachers and students integrates with the timber industry at a local, regional, or national scale. In others, the integration is with a specific forest or woodland. In others, it is not the forest itself but the means of embedding the architectural interventions undertaken in their local context. The locality of these initiatives is central to understanding their way of working, their concerns, and their interests.

Sourcing locally produced timber is one way to enhance its sustainability: a shortened journey and efficient logistics can significantly reduce the environmental impact of transportation and handling. Being familiar with the processes behind timber production and its impact on landscape, environment, economy, and society can be linked with a more conscious use of the material (and understanding its necessary cycles and methods of maintenance).

Amongst the cases we have analysed, the more research-oriented ones are part of a wider economic ecosystem where forestry and timber production are central. Three of our case studies (Gramazio Kohler Research, IBOIS in Lausanne, and ICD/ITKE in Stuttgart) are all located in proximity to the forests and forest industries of the vast trans-border Alpine region. The Wood Program at Aalto University is in the Nordic region where forestry is a key economic sector. In the United Kingdom, both the AA's Hooke Park campus and the Studio in the Woods are in specific forests with which they have established a symbiotic relationship. Other examples in this book, such as the UoN's Design+Build studio, RUF, and the Insitu Project, work nationally or internationally but engage with the locality of the wood they use through a critical awareness of the availability of material, social, and cultural resources.

6.3 SOCIAL ECOLOGIES

For many of the architects, teachers, and students involved in the initiatives described in this book, the choice of wood as a building material can be ascribed principally to a concern for environmental sustainability. But many of these projects are also defined by an equal or greater concern for social (and societal) sustainability.

Wood is produced in forests, and the forests that produce wood are an important part of human and non-human landscapes. The European Landscape Convention (ELC) identifies how the sustainable preservation of landscape leads to a harmonious coexistence between local communities and their economies, which contributes to the physical shaping of their life environment (Council of Europe, 2000).[1] It is particularly important to consider the relationship between landscape and society in terms of wood production: as Jane Hutton underlines in her book *Reciprocal Landscapes* (2018), the extraction of a material from a place has deep social and aesthetic consequences for its landscape, even if that landscape is uninhabited, sparsely populated, or far away from the place in which the material is eventually used.[2]

The forest industry has a significant impact on the landscapes that produce its raw material. Forests present a large physical presence: large enough to envelope human activity and human society. When forests are actively managed, whether at a small scale or at a larger industrial scale, their transformation through logging can be negatively perceived by local communities. Choosing to use wood as a building material can have significant social and societal implications, impacting not only landscapes but also local economies. The construction of a building with public elements also has implications on its surroundings, changing the lives of the communities that engage with it. The construction of such a building using timber adds a dimension that engages the building user with the landscape that was changed by the production of the building.

The cases we have analysed engage with social aspects in different ways and with different degrees of trans-national activity. The primary goal of the UoN's Design+Build Studio is to realise semi-public buildings (mainly childcare facilities) for underserved communities. These buildings must be resource efficient, high quality, and durable. In China, the projects realised by Insitu are inspired by a holistic understanding of sustainability, which translates into a strategy in which the use of wood is linked to the reactivation of local traditional knowledge. This is aimed at recreating a local community identity and at the same time building a community centre and rebuilding local culture.

Social sustainability may not appear to be a concern of the initiatives that locate themselves deep in the forest. But some, like the AA's Hooke Park, provide an opportunity to realise a fully circular production process that can illuminate more sustainable strategies for design and construction. Such work may appear to be inward looking (as opposed to outward), but it can provide a platform that allows the academic institution to develop as a stakeholder with a distinct ethical positioning, which can better serve it when engaging with communities.

6.4 MATERIAL UNDERSTANDING

Wood has unique physical properties which are essential to understanding its capacity for sustainable use in construction. The case studies in this book illustrate a variety of ways of understanding the properties of wood as a building material. In each case, a deep understanding of the material at hand helps teachers and students to achieve their goals.

Figure 6.3 Prototyping experimental joints in the workshop of Aalto University's Wood Program. Photo: Juri Kikuchi.

Some initiatives (such as ICD/ITKE, Hooke Park, IBOIS (the Laboratory for Timber Constructions at the École Polytechnique Fédérale de Lausanne) and the Aalto Wood Program) engage in structural experimentation that tests or extends the perceived limits of the physical properties of wood. In many cases, these experiments are supported by the application of digital technologies, in surveying, harvesting, manufacturing, or construction.

On a microscopic scale, laboratory-based technology can be used to study the molecular behaviour of wood fibres in organic and manufactured wood products, allowing architects to not only control but also programme the form and behaviour of wood elements. On a larger scale, technology can be used in the forest and on the building site to explore the organic language of naturally occurring geometries in trees. Understanding how different species of trees grow in different contexts and settings can enable students of architecture to understand how wood can be harvested from the forest and used in construction with minimum processing and waste.

As discussed previously in this chapter, access to technology should not be considered a barrier to achieving this kind of work. Traditional craft techniques, such as cruck-framing in timber building, can give us insight on how to reimagine traditional building techniques in contemporary projects. In China, the Insitu Project has demonstrated a sustained strategy for the revitalisation of traditional building techniques that reflect both the human and non-human environment, reviving traditional craftsmanship and responding to the different material characteristics of the locally available tree species. Looking at traditional building methods with modern eyes allows architects to understand the motivation behind certain material choices where, for example, certain trees are used for specific structural elements. They link material understanding to the local culture, preserving it and making it part of a wider strategy for sustainability.

6.5 ECOLOGICAL UNDERSTANDING

In Chapter 3, we argued that wood has a transparent ecology: a set of characteristics that allow for an easier comprehension of wood's ecology compared to that of other fossil- or mineral-based building materials. Through the case studies, we have considered a variety of approaches aimed at helping students understand the

Figure 6.4 The forest landscape of Tavelsjöberget, Västerbotten county, Sweden in mid-winter. Photo: Francesco Camilli.

transparent ecology of wood. We believe that these case studies demonstrate innovative methods of teaching architecture that are more effective than those used in normative, mainstream architectural education.

Ecologies are interconnected networks of ever-changing relations. Understanding the ecology of wood as a building material means understanding the complex ecologies that produce it and transform it from a raw material into a manufactured product. In some of our case studies, students of architecture begin their engagement with wood before it has been harvested from the forests that produced it. While a short field trip to a forest is not enough to begin to understand these ecologies, the initiatives discussed in this book offer a variety of different ways of challenging architecture students to think not only about where wood comes from but also about how wood that does not fit the normal standards of industrial forestry can nonetheless be used productively. On two very different scales, the AA's Hooke Park campus and the weekend-long Studio in the Woods programmes are both concerned with establishing a direct relationship with their immediate surroundings, exploring how the by-products or waste materials of industrial forestry can be productively used.

Other initiatives, such as the case studies from Switzerland, Germany, and Finland (Gramazio Kohler Research, ICD/ITKE, IBOIS, and the Aalto Wood Program), may not incorporate such a direct student engagement with the forest, but they demonstrate long-term (multi-year) engagement with the regional economic and political environment. Close connections to industrial actors provide access to materials, know-how, and projects that enrich the educational experience. These initiatives, especially when oriented towards both research and education, explore the ecological properties of wood on a wide range of dimensional scales, from the molecular to the urban (Ibañez, Hutton, and Moe, 2019).

The more socially oriented case studies are not only concerned with environmental sustainability, but also with human and social sustainability. In building or renovating community facilities in rural China, the Insitu Project places a high value on the contributing role of established craft and production cultures. These cultures connect and strengthen the relationship between students and community members and their local environment. That connection can – if special effort is made – also contribute to a better understanding of the roles of non-human stakeholders (including animals and plant life) in the expanded ecology of the building.

Each of these different approaches shares a commitment to allowing students of architecture to work hands-on with wood. We argue that they give students an unparalleled opportunity to engage with wood's complex yet transparent ecology.

Notes

1 See https://coe.int/en/web/landscape.
2 *Forest Trail* is a small research project under way at Umeå School of Architecture (2024–25), with support from UmArts. The project will follow a tree all the way from the forest that produced it to the building in which some of its wood will eventually be used.

Bibliography

Council of Europe. 2000. *European Landscape Convention*. Florence: Council of Europe.

Hutton, Jane Elizabeth. 2018. *Reciprocal Landscapes: Tracing Materials Between New York City and Beyond*. 1st edition. London: Routledge.

Ibañez, Daniel, Jane Hutton, and Kiel Moe. 2019. *Wood Urbanism, From the Molecular to the Territorial*. Barcelona: Actar Publishers.

Smith, R. Leo and Stuart L. Pimm. 'Ecology.' Encyclopedia Britannica. Accessed 22 February 2024. See https://www.britannica.com/science/ecology

Gramazio Kohler Research, ETH Zürich, Switzerland

DOI: 10.4324/9781003428930-13

Figure CS7.1 Gradual Assemblies Pavilion, Istituto Svizzero, Rome, Gramazio Kohler Research, 2018. Photo: Valerie Bennett.

Gramazio Kohler Research (GKR) is a research unit at ETH Zürich (the Swiss Federal Institute of Technology in Zürich). ETH is a public university founded in 1854 and modelled on the *École Polytechnique* in Paris, with sixteen departments focusing on sciences, technology, engineering, and mathematics.

Like the collaboration between the Institute for Computational Design (ICD) and the Institute of Building Structures and Structural Design (ITKE) discussed in Case Study 3, GKR is a research unit, but since 2006 it has also provided a platform for workshops, seminars, electives, and core courses to students enrolled in the Department of Architecture. The research undertaken examines changes in architectural production resulting from new digital manufacturing techniques, combining data and material to interrogate the consequences for architectural design. Attention is paid not only to the possibilities of new techniques for digital fabrication but also to design's relationship to production and the resulting influences on architectural aesthetics. The aim of teaching activities is to produce graduates who are experts in digital fabrication, computation, and robotic fabrication.

Figure CS7.2 Gradual Assemblies Pavilion, Istituto Svizzero, Rome, Gramazio Kohler Research, 2018. Photo: Gramazio Kohler Research, ETH Zurich.

Figure CS7.3 Plan drawing of the Gradual Assemblies Pavilion, Istituto Svizzero, Rome, Gramazio Kohler Research, 2018. Photo: Gramazio Kohler Research, ETH Zurich.

Switzerland has a long history of building in wood, with a highly developed ecosystem of forestry, construction, and engineering companies. The supply chain supporting GKR is dependable and their good relationships with established partners are credited with contributing to their research and teaching success.

GKR designs, tests, fabricates, and builds with a variety of materials, not only with wood. However, their work offers us a practical prototype of research-informed pedagogy. The relative merits of wood (easier to work with, cleaner to work with, and requiring less material preparation) are acknowledged. Hannes Mayer, who was a Senior Researcher at GKR between 2016 and 2022, argues that 'timber is something you can cut into the size of a student's arm span… which is why it lends itself well for collaborative projects.'

GKR use wood because of its affordability, light weight, and good structural performance. Robotic fabrication is used to complete structures that are more complex than those that could be built by hand. Mayer reflects on how, in the development of engineered timber, 'we tried to turn timber into a concrete beam. We came up with glulam and cross-laminated materials to avoid any uncertainty or uncontrollable behaviour that is introduced by its being a natural material.' GKR's projects use computational control and robotic fabrication to work creatively with wood in a way that does not resist or suppress the naturally occurring tolerances and behaviours of each piece.

To give two examples, a pavilion for the *Istituto Svizzero* in Rome and an indoor installation for the Victoria and Albert Museum in Dundee were completed by Masters students on the Advanced Studies in Architecture and Digital Fabrication programme. Using the same structural principle, short spruce slats were held in place by beechwood dowels to create a lattice. Computational design was used to precisely design the arrangement of both the slats and the dowels, which could then be translated accurately using robotic fabrication. The beech dowels were kiln-dried and then rehydrated so that they expanded and locked into place. The projects allowed students to experiment with and understand the heterogeneous qualities of different kinds of wood – designing, fabricating, and constructing in a way that used the natural

Figure CS7.4 Gradual Assemblies Pavilion, Istituto Svizzero, Rome, Gramazio Kohler Research, 2018. Photo: Martina Cirese.

Figure CS7.5 Gradual Assemblies Pavilion, Istituto Svizzero, Rome, Gramazio Kohler Research, 2018. Photo: Martina Cirese.

behaviours of the materials to minimise the use of any epoxies, adhesives, or metal fixings.

Reuse of materials is an important strategy in this approach to design. Eliminating non-wood materials increases the possibilities for dismantling and deconstruction or, in the worst-case scenario, recycling into other materials or energy. Amongst the case studies described in this book are pavilions assembled on an annual basis by schools of architecture. Mayer prompts us to 'think about how much waste education actually produces because if every faculty builds a pavilion every year, it's quite a bit of waste.'

Case Study 8

Insitu Project, Hong Kong Polytechnic University and Shenzen University, China

DOI: 10.4324/9781003428930-14

Figure CS8.1 Miaoxia Community Guesthouse (background), 2018, and Community Kitchen (foreground), 2016. Part of the Miaoxia Community Projects. Photo: Insitu Project.

Most of the literature relating to design-build projects in architectural education is derived from initiatives in North America and Europe. Yet there are precedents for hands-on work with wood in architectural education elsewhere in the world, including those set in regional contexts that bring new complexities to matters of carpentry, craft, and working with wood.

The Insitu Project is a research-by-design platform established by Peter Hasdell and Kuo Jze Yi. The initiative is a non-profit association based at the two universities where Hasdell and Yi teach: the School of Design at Hong Kong Polytechnic University (PolyU) and the School of Architecture & Urban Planning at Shenzhen University (SZU). A number of collaborations have taken place between the Insitu Project and other departments, institutions, non-governmental organisations (NGO) and community groups on various sites in Guangdong, Henan, Hong Kong, Hunan, Shandong, Shanxi and Sichuan.

Operating in one of the world's most rapidly urbanising and developing countries, the main goal is to promote sustainable development in rural areas affected by demographic decline, and where the local culture has been weakened by the major economic transformations of the country. Insitu conducts site-specific projects: sometimes they realise small buildings, but their work also includes spatial interventions, site and context investigations, spatial planning, social enterprise formation, and workshops.

To do this, Insitu projects address sustainability through three crucial and complementary aspects: materiality, society, and economy. Material economy concerns the use of locally available materials that are connected to local culture and traditions and therefore allow them to rely on local knowledge; societal sustainability prioritises the realisation of structures that can support the cohesion and common goals of communities; and economic sustainability is about supporting local economies that can make the projects economically viable in the long run. These aspects are incorporated through dialogue with local stakeholders and through understanding the needs, the culture, and the skills of a particular place and how these are linked to the locally available resources and materials.

Figure CS8.2 Miaoxia Community Kitchen under construction, 2016. Photo: Insitu Project.

Consequently, while hardwood is not a specific focus of the Insitu project, the ancient Chinese tradition of wooden construction (particularly evident in the elaborate roof structures of historical buildings) makes timber one of the main materials used in their work, acting as a bridge between the traditional knowledge of local residents and the experimental approach of students. These traditional building methods are deeply connected to the physical context, relying on locally available materials and their peculiar properties. In the words of Hasdell, 'this is part of the sustainable cycle in my view because it is how the human or cultural skills actually develop at the same time as the harvesting of this renewable resource.'

In the contexts where Insitu works, these traditions, replaced by cheap, industrial technologies and affected by decades of modernisation policies, are usually devalued and almost forgotten. Therefore, a key element of Insitu's sustainability strategy is to revitalise them, thus favouring a renewed connection between the inhabitants and their living environments. In their interview with the authors, Hasdell explained how:

> these things have been disrupted over the past sixty or seventy years of Communist rule and rural policy changes … we are not claiming to put [them] back in place but just to understand some of these aspects and understand where the material cycle fits in … the local environment.

Insitu couples this rediscovery of traditional building methods with the innovative design approach developed by researchers and students. Through this strategy, they realise buildings that reinterpret tradition in a modern way, thus creating contemporary structures that are in continuity with traditional ones.

In their projects, students and local stakeholders can learn from one another, allowing the former to be exposed to traditional woodworking crafts and skills not taught in the academy and the latter to learn about the techniques of contemporary architectural research and practice. A careful understanding of the local conditions, materials, and skills is needed to identify and promote sustainable practices.

The structure of study curriculums does not allow students to be involved in the entire design and realisation process. Instead, they engage in thematic workshops where they can focus on specific aspects of the project, performing local surveys and dialogue sessions together with design activities. Working with stakeholders who are not familiar with drawings, 1:10 and 1:50 models have become a signature methodology for discussing and developing designs collaboratively.

Case Study 9

Rural Urban Framework, University of Hong Kong, China

DOI: 10.4324/9781003428930-15

Figure CS9.1 GerHub, Ulaanbaatar, Mongolia, Rural Urban Framework, 2019. Photo: James Benedict Brown.

Rural Urban Framework (RUF) is a research and design collaboration established in 2005 by Joshua Bolchover and John Lin, two assistant professors at the University of Hong Kong. Bolchover and Lin were motivated by the Chinese government's publication of a plan to urbanise half of China's 700 million rural citizens by 2030.

RUF operates as a non-profit organisation of the Faculty of Architecture at The University of Hong Kong; it collaborates with charities, private donors, governmental departments, and universities. Since its establishment, RUF has completed more than 15 projects in China, Mongolia, and Nepal, including schools, community centres, hospitals, village houses, bridges, and incremental planning strategies. Through these interventions, RUF has also conducted research into the links between social, economic, and political processes and the physical transformation of each village. As RUF does not work exclusively with timber, we did not initially identify their work in our search for case studies. However, when we had the opportunity to study their 2019

GERhub project in Ulaanbaatar, Mongolia, it quickly became apparent that their work had relevance to our study.

Since the nineties, the city of Ulaanbaatar has experienced an exponential population growth. Many Mongolians abandoned their traditional nomadic life to move to the city. The housing stock of the city has not absorbed this growth, favouring the growth of informal residential areas made of *gers*, traditional light-weight buildings typically used by Mongolian nomadic communities as temporary, movable dwellings. These areas, where the air is heavily polluted by the extensive burning of coal for domestic heating and cooking, lack not only basic urban infrastructures but also social spaces.

The Ger Innovation Hub (GIH) project aims to address this lack of social infrastructure: GerHub is a non-governmental organisation (NGO) operating in the Ger districts of Ulaanbaatar, trying to foster just and sustainable development through innovative and creative solutions. It has collaborated with RUF for almost a decade on initiatives aimed at improving living conditions of *ger* districts in Mongolia. In the beginning, these focused mostly on basic technical infrastructure. However, the need to address social issues in the *ger* districts quickly emerged, inspiring the conception of a building that could host social activities organised by GerHub.

RUF undertook this commission as a design-build project, involving students from two Hong Kong-based higher education institutions, the University of Hong Kong and the Hong Kong University School of Professional and Continuing Education, ensuring a more multidisciplinary setting. To oversee the construction, they also involved the local NGO Eco Nogoolin Ger Community Social Enterprise (Ecotown), which specialises in sustainable, affordable building technologies, and has strong links to the area and its community.

Together they have realised a building that, while being inspired by the nomadic *ger* in form, addresses some of the limitations of this building type when used as a dwelling in an urban context. The traditional *ger* is a circular tent covered and heated by a stove fuelled by wood or coal. It is conceived for nomadic life, so it does not have any hygiene facilities; when present, these are often situated in poorly built, more traditional houses next to the *ger*. This, together with the poor ventilation and the elementary heating

Figure CS9.2 The intermediate space between the internal room and the external polycarbonate shell. GerHub, Ulaanbaatar, Mongolia, Rural Urban Framework, 2019. Photo: James Benedict Brown.

system, generates an unhealthy environment both inside and outside, exposing these populations to high levels of pollution.

The GIH is composed of an inner core built of locally manufactured mud-bricks walls, and an external polycarbonate shell. This creates an internal room, which is protected from the external climate, and an intermediate space between the core and the external layer. Here, the passive solar gains generate milder temperatures, allowing for outdoor activities during the frigid winters of Ulaanbaatar. This configuration was made possible by the use of timber: the structural concept based on the use of trusses allowed for a double roof to be built, where the external layer is anchored to the top of the truss and the internal to the bottom. The wood used to build the Ger Innovation Hub was sourced from either Mongolia or Russia, making it cheap and easily available.

Timber also had a pedagogical role: students engaged in experiments in the university workshops – for example, by realising a 1:5 model of the structure and 1:1 mock-ups of joints – and in the realisation of the building onsite. Moving from workshop to site challenged the students to respond to the complexity of the context: this was not just limited to unexpected building issues but also to the social environment they were confronted with. While on site, students conducted surveys as a way of engaging with the residents and determining their hopes, aspirations, and needs.

The project started in 2018 with focus groups aimed at planning events and engaging the community. During the construction process, several meetings and social events were organised to inform the community and create engagement. After the external layer of the building was completed, sensors were installed to monitor its environmental performance. The building was completed in 2020.

Close to the GIH, GerHub and RUF have also realised a Demo Ger, a model building that shows ways to make *gers* healthier and more liveable. This building facilitated the organisation of several workshops in which people were trained on how to make an effective DIY insulation, how to use clean firing methods in their houses, the health impact of air quality, and how to manage household finances.

Chapter 7
The Politics of Wood

An architect's choice to use wood as material in a building has consequences for human and non-human stakeholders who share a complex web of relationships around wood and forests. Countless negotiations of interests take place in this network, challenging its ecological balance continuously. This dynamic can be compared to a political arena in which different interests confront, contrast, and balance each other, eventually coming to a common equilibrium.

The word 'politics' comes from the Ancient Greek πολιτικά, meaning the affairs of cities, in which the coexistence of large amounts of people with their own interests needed to reach an equilibrium to ensure the cities' prosperity. Politics is therefore a concept that mostly refers to human relations and actions, and to how they can converge to achieve greater results than those that could be achieved by individuals alone.

Applying a human concept like politics to a non-human material like wood might seem like an anthropocentric act. Nevertheless, we already discussed how wood is the product of a complex forest ecology, and how an ecology is a web of relations between different actors in the environment.

Forests and cities alike can, therefore, be considered as the products of continuous negotiations between the interests of different actors. Politics is a relevant concept to apply to the forest, able to generate reflections on the complex impacts that the use of wood can have on different stakeholders involved in its life cycle.

DOI: 10.4324/9781003428930-16

Even if we acknowledge their existence and importance, this chapter will not delve into ideological aspects of politics. Instead, it will focus on the kinds of relations that can be generated around the practice of building with wood in an educational context, referring to a non-anthropocentric understanding of reality that gives human and non-human actors the same level of legitimacy and agency (Latour, 2005; Harman, 2017). This brings aspects of regenerative design into play: approaches that work with or mimic the natural ecosystems in the environment to make a positive contribution to the health of other living species.

By understanding politics as the way all these actors interact and considering what triggers these interactions and the consequences they can generate, this chapter will recognise stakeholders' impact on communities and knowledge as a resource. Finally, this chapter will reflect on how wood can generate collaboration.

7.1 STAKEHOLDERS OF WOOD

Buildings and building materials have stakeholders. Like any other material, wood is a building material that binds together many disparate and different stakeholders, including forest owners, machine operators, truck drivers, sawmill operators, retailers, engineers, architects, construction labourers, clients, civil servants, politicians, and the end-users of the building.[1] In our case studies we have seen how – when wood is used in hands-on teaching and learning initiatives in architectural education – the list of stakeholders includes teachers and students, as well as live project or design-build project clients, their current and future users, local communities, funding providers, and educational institutions. In many of the projects we have shared in this book, stakeholder mapping is a productive and valuable collaborative exercise for helping students to comprehend the complexity and consequences of their actions.

Writing this book from the perspective of architectural education, we tend to regard students as the primary stakeholders in these initiatives. The role of the teacher is to provide them with educational insights that can support their educational development and future careers. As we argue in Chapter 3, the accessibility and versatility of timber make it easier for students to work hands-on with it. But through hands-on experience of construction, students of architecture can potentially come closer to considering more directly the clients and future users of a building. These stakeholders are often missing or relegated to silent simulation in a normative student design project. In some of the examples included in this book, clients and end-users are actively involved in the collaborative realisation of a building alongside students and teachers. Working with both a non-governmental organisation (NGO) and the headteachers of small crèches in South Africa, the University of Nottingham's (UoN) Design+Build Studio has demonstrated a commitment to not just working with but actively involving two very distinct stakeholders: the client (the NGO) and the end-user (the teacher who will one day take ownership of and responsibility for the finished building). For stakeholders in the Design+Build project, the handover of the building at the end of the construction has become a vital transition: the building is no longer the property of the students, teachers, or university. The end-user, in their opinion, must be supported to take ownership of it and not become dependent on those who built it.

Stakeholders in the forest industries, wood production, and logistics chain play an active role in providing materials, tools, and capacity for the realisation of these student projects. Local communities and governments are affected by spatial transformations generated by the projects and in the territorial transformations resulting in the use of the forest as a source of building materials. They engage with these dynamics at different levels, from the aesthetic or cultural appreciation of forests to the active involvement in local economies, and are therefore interested in taking part in these processes.

But adopting an ecological approach to realising buildings also implies that non-human actors can be considered as fully legitimate stakeholders. As we suggest in Chapter 2, the forest can also be considered an active stakeholder, since timber is a product of its complex ecology. Considering the forest as a stakeholder opens up further reflections on how to understand ecology and sustainability. The forest is itself a system of different stakeholders, not limited to trees, but also including any other kind of animate or inanimate being that populates it and contributes to the equilibrium of its ecosystem. At the Architectural

Association's (AA) Hooke Park campus, for example, teachers and students do not attempt to increase the productive capacity of the forest; instead they restrict their use to the volume of material that can be sustained by the ecological cycle of the forests. This suggests a model of approaching forests not as passive sources from which materials are extracted, but as active material providers whose capacity needs to be taken into account and respected.

Zooming out from the material to the environments that produce it, forests are embedded in larger territories. According to the European Landscape Convention (Council of Europe 2000), the sustainable relationship between communities and their territories is what generates and preserves landscapes. Applying this vision to the forest can help us to understand how this environment is deeply affected by its relationship with the different stakeholders that populate it and, at the same time, how it shapes the life of these communities (Magnaghi, 2020).

Stakeholders contribute their vision, their needs, and their knowledge to the process of making a building. In exchange, they get a voice in the process, they benefit from the improvements brought about by transformation, and they enjoy the opportunity to exchange knowledge. These stakeholders express their needs and aspirations regarding the transformation of their environment: clients and future users are interested in new spaces that can comfortably host their activities; local communities pursue the improvement of their living conditions; local governments promote the economic and social development of their communities; and forestry industries seek profitable returns on their investments through increased efficiencies and productivity.

By engaging with the different stakeholders involved in their projects, students can experience the complex biological, social, economic, and cultural implications of the use of timber in construction. An example of this can be found in the work of the Insitu Project in China, their dialogue with local communities allows for the revitalisation of traditional building techniques that use local materials. Students engaging in these activities can observe the connections between local cultures and their living environments and understand how these connections can contribute to sustainable local practices. By mapping stakeholders and their contributions onto design build projects such as those

Practising Wood in Architecture

studied in this book, students of architecture are placed in a unique position that allows them to grow a deeper understanding of the construction process. While learning is the main goal for students (about the characteristics of the ecology of the forest, the design of a building, and so forth), other stakeholders can then learn from their new insights. Teachers and scholars have the opportunity to test their research findings and approaches. Clients can be stimulated by the ideas proposed by the designers, gaining new perspectives on their vision for spatial transformation. In a similar way, future users and communities can learn about how spatial transformations can affect their lives and how their relation to timber and the forest can evolve in a sustainable way. Local governments can gain knowledge about their territory and communities, while having an opportunity to experiment with social cohesion. Industry can learn about innovative methods that can be integrated into their working practices.

For these reasons, we argue that the choice of wood in architectural construction can be understood as a political act. More than any other material, wood allows those who design and make buildings to connect with stakeholders at a wide and even territorial scale. Mapping and naming the human and non-human stakeholders of wood is a critical step in the hands-on use of wood in architectural education.

7.2 THE SOCIAL IMPACT OF WOOD

We have seen how the decision to use timber in construction affects large and complex networks of human and non-human stakeholders. In the educational context, acknowledging the social effects of material choices becomes crucially important for the formation of responsible practitioners. The social impact of the case studies presented in this book can become tangible at different scales, from affecting the spatial transformation of entire communities to the deeply personal educational experience of future practitioners.

At a local level, realising timber buildings in collaboration with local communities can provide those communities with new spaces that can promote social cohesion. The use of timber can bring together a diverse range of people in a building process, promoting social cohesion around common interests. Experimenting with new building techniques can contribute to generating new economies based on innovative technical knowledge. The use of timber is also linked to forestry practices that contribute to shaping landscapes recognised by local communities. Landscape is one element of a community's identity and its modifications have direct impacts on the local social fabric of a community (Council of Europe, 2000; Hutton, 2018).

Owing to the scale of their operations, forest industries can have a significant impact on local environments and communities. Forests are an important aesthetic presence that communities continuously perceive as part of their living environment. Aesthetic changes generated by human activities can therefore also affect the daily experience of the inhabitants of a community. Throughout the process of writing this book we have turned to the forests of northern Sweden for inspiration. Here, we observe the impact that forest industries have on ecosystems and the lives of the different human and non-human stakeholders that populate the forest (Kivinen et al., 2010). In Sápmi, the homelands of the indigenous Sámi people that span Norway, Sweden, Finland and Russia, the impact of industrial forestry also affects reindeer herding. Even though reindeer herding is just one small part of contemporary Sámi culture, this conflict contributes to a problematic social divide and perpetuates a long history of colonial exploitation. Participative and inclusive forest management could be key in promoting better social relations and preserving Sámi heritage.

Industrial forestry, even when marketed as sustainable, conflicts with the conservation of biodiverse natural environments. The practice of clear-cutting entire plots and arguing for improved carbon compensation through the replanting of new trees has a deep impact on the forest ecosystem, exposing the terrain that hosts lichen and other smaller, yet fundamental, living species. Reindeer, for example, which feed on these plants, are deeply affected by these practices, losing vital sources of nutrition in cold winter months. It also remains a matter of intense debate whether monocultural forest plantations, with their lower levels of biodiversity, can have the same carbon capture and storage potential as more biodiverse old-growth forest.

In our case studies we have found various different means of addressing the social impacts of building with wood. The buildings realised by the Insitu Project make an important contribution to improving the conditions of local communities. The participatory design process brings people together, encouraging them to cooperate on a common project; the rediscovery and updating of traditional building techniques strengthens and preserves the communities' cultural identity; the reintroduction of the use of local materials can generate new sustainable economies; and the new buildings provide social services and opportunities for economic development. At the AA's Hooke Park campus, the attempt to build a community of learning, research, and practice rooted in the forest represents a push towards the institution acting as a stakeholder with an ethical stance. Hooke Park and the practices undertaken there set an example for a sustainable, and one day circular, relationship with the forest. And like so many design-build projects built in wood, the GerHub built by Rural Urban Framework (RUF) in Ulaanbaatar sought to involve as many local stakeholders as possible in its construction, creating a community centre that promotes sociability in an otherwise alienating urban environment.

On a larger scale, we have seen how promoting the sustainable use of timber as a building material can be seen as one of the strategies through which climate neutrality can be pursued in the construction industry. The focus on sustainable timber and a more ecologically aware approach to forestry can promote new, more caring ways of interpreting economic development. Innovations in management, logistics, and governance can improve collaboration amongst different stakeholders pursuing more sustainable economies. The involvement of many different stakeholders in this collective effort has deep social implications that question global economic paradigms.

Proposing new modes of industrial cooperation between forestry, timber industry, public administrators, designers, etc., can be seen as a way to positively affect the societal (political) issue of sustainability. To give one example, in Germany the Institute for Computational Design (ICD) and the Institute of Building Structures and Structural Design (ITKE) promote an approach to the use of timber that is explicitly oriented towards material economy, aiming to demystify the use of timber and to promote innovation within the industry. The buildings realised demonstrate new approaches to building technology, promoting more sustainable industrial and economic ecosystems. In Switzerland, IBOIS (the Laboratory for Timber Constructions at the École Polytechnique Fédérale de Lausanne) has pursued collaboration between local industries, decision makers, and communities who work around wood and share the goal of reinforcing local sustainable logistics; this is their way of promoting sustainability at a systemic level. Finally, in Finland, the Aalto Wood Program operates in the context of a national push towards the development of timber building technology and the promotion of timber as a sustainable building material. The projects we have studied endorse a culture of sustainable timber industry and buildings, aiming to transform public spaces and to improve the quality of life of residents.

7.3 LOCAL KNOWLEDGE OF WOOD

Trees cast their roots deep into the ground, searching for nutrients and securing themselves against climatic events. We do not intend to invoke any kind of cliché, but wood is similarly rooted in its local culture. Local knowledge about and understanding of forest management and the use of wood can contribute positively to the creation and maintenance of sustainable building processes.

As a material used in the construction of buildings and the manufacture of commodities, wood connects human communities with their territories. Forests are an integral part of the life of the communities that are in contact with them. Having established relations with forests and a knowledge of their behaviour, communities develop a deep understanding of forest ecologies. They take care of them while also taking resources from them. In this way forests are shaped by communities and influence their lives. This mutual dependence is what generates landscape (Magnaghi, 2000) and can produce in communities an understanding of their living environment that is deeply rooted in its ecology.

In the case studies presented in this book, this local knowledge manifests itself in different forms. Many of the case studies we present here are situated in areas that have strong traditions of using timber as a building material because of their proximity to forests. To give several European examples, OIS, Gramazio Kohler Research, and ICD/ITKE are all situated in the broader Alpine region, where forests are an important part of the landscape and economy. They rely on a long tradition of timber building that has gone through the advanced industrialisation process that characterizes this part of Europe. The advanced technologies that they use can be developed because the technological level of the local timber industry is already one of the most modern in the world. In this way, academic institutions can collaborate with local industries in researching cutting-edge technologies that are rooted in the local landscape and tradition.

Local knowledge can also be interpreted as the development of knowledge of a specific site: understanding a specific environment and its ecology can contribute to the exploration of more sustainable ways to interact with them. In England, this can be seen in the work of the AA's Hooke Park, where students and teachers explore the forest surrounding their campus and the materials that it produces. By knowing what tree species are present and by paying attention to the historical dynamics that created the forest, explorations of ecological cycles that can support timber construction are possible. The buildings constructed here are generally small and bespoke, but they provide operational examples for architects to take inspiration from.

While the industrialisation of the building industry has seen timber construction evolve away from bespoke buildings that are adapted to the available materials and towards more standardised, commoditised products and technologies, the rediscovery of local traditional techniques and their integration with new technologies and modern design can help re-establish the ecological relationship between a community and its living environment. By involving local communities in educational architecture projects, traditional building technologies can emerge and be explored by students, who can learn from them and use them in their design process to meet modern needs. Wood can be reintroduced as a local material, producing buildings that are better integrated both in the local environment and in the local culture. An example of this dynamic can be seen in the work of the Insitu Project, where contemporary buildings are inspired by traditional building techniques.

Although the GerHub in Ulaanbaatar is a strikingly modern building that uses modern building components and materials, its design and configuration is inspired by the traditional Mongolian *ger* (a circular tent or nomadic building). A community centre that is integrated into the local built environment and provides space for social activities, it was realised by a team comprising students and local residents who connected architectural innovation with traditional building methods.

To work with wood is to be open to the possibility of long traditions of craftsmanship and design. When students of architecture get to work hands-on with wood, there is an opportunity to engage with local culture and knowledge that is often excluded from the academy. Naming, integrating, and even sometimes updating this knowledge is a viable and meaningful strategy to engage communities in the effort towards more sustainable building and territorial management practices.

7.4 WOOD AS A COLLABORATIVE MATERIAL

Like the politics of society, the politics of wood encompass the decisions, strategies, tactics, and processes of groups of people. Politics demands debate, negotiation, compromise, and the exercise of authority, and we argue that wood can be understood not only as a political material, but also as an inherently collaborative material. The complexity of even the simplest building project normally requires collaboration among different actors. Wood can be particularly suited to promote this collaboration. With its transparent ecology, the life cycle of wood as an organic material can engage multiple stakeholders in negotiation, design, construction, and the procurement of resources.

The sustainable management of any material's life cycle requires the cooperation of as many relevant stakeholders as possible. In an educational context, the collaborative characteristics of wood can help to develop in young professionals a sensitivity towards the social responsibility that is linked to the design practice. Giving students the opportunity to confront a simple yet complete participatory design process that includes both co-design and collaborative building can be a significant experience in their educational pathway towards becoming a responsible citizen architect (Santanicchia, 2020).

Wood and timber can facilitate collaboration among different stakeholders in different ways. For example, the versatility (see Chapter 3, section 3.3) and accessibility (Chapter 3, section 3.4) of wood make it an ideal material for cooperation between groups of people with different backgrounds and expertise. Two students of average physical ability can pick up, carry, and position a variety of standard structural timber elements. The strength, flexibility, and forgiving nature of wood facilitates the experience of learning to design and build through actual construction.

The variety of applications for which wood and timber are suitable make them a preferred choice when realising collaborative projects. Collaborative projects require effectiveness and simplicity in construction, which timber can easily provide. The simplicity of handling wood can remove the obstacle of technical training, making it possible for more people to be involved in the building process. This is particularly favourable in an educational context. For example, students can work together with residents and future users in an easier way when they do not need to be trained in the handling and assembly of the material. This emerges in the work of the UoN's Design+Build Studio which, during its many years of activity, has built relationships

Figure 7.1 Wood facilitates collaboration due to its strength and lightness. Two students of average physical ability can pick up, carry, and position a variety of standard structural timber elements. Photo: Thabang Radebe.

with local actors in the fields of construction and primary education. Students benefit from the support of a local social and industrial infrastructure, but this can also become a challenge for students who have to adapt to the traditions, habits, capacities, needs, and wishes of a network of local stakeholders.

Collaboration with local (non-academic and non-architectural) stakeholders can become a necessary step to realise many projects. For example, when working in remote environments, having a strong network of local stakeholders that can provide groundworks and assist with administrative processes is crucial for the positive outcome of the project. Dialogue with local stakeholders can also help to address the continuous process of deconstructing colonial relationships (such as when students are instructed to design for an environment that they are not familiar with).

Collaboration is also a vital step towards using timber sustainably. A well-functioning logistical process involving all actors from forest management to designers can be crucial in the correct management of forests and wood. Industrial, technical, administrative, and local stakeholders can cooperate to establish local, integrated, and sustainable building processes. Different disciplines like forest management, biology, architecture, and engineering can collaborate in research and practical experimentation.

Academia can have a crucial role in promoting cooperation as it seeks to involve relevant stakeholders to promote its research and make it impactful. Involving relevant actors from the forestry and building industries is crucial for testing new technologies in the field and promoting their application. Higher education and research institutions can also contribute to the creation of local industrial ecosystems that can both incentivise and take advantage of technical innovations. Many of the case studies we have analysed illustrate this dynamic. In Germany, ICD/ITKE work in a region in which forestry is an important part of the local economy. Their work aims to promote technological development by transferring the knowledge they produce to local industries. In a similar way, in Switzerland, IBOIS enters into dialogue with both local decision-makers and industrial actors to promote cooperation around forestry and timber construction on a local level. Finally, in Finland, the Aalto Wood Program collaborates with local decision-makers to propose projects that can have a positive impact on communities, realising quality public space and buildings that are open to social use.

Other initiatives have a more pedagogical interpretation of collaboration: the short weekend design-build projects undertaken by the Studio in the Woods bring together students and specialists from different disciplines to intensively work together for a short period of time. Wood's ease of use facilitates this quick collaboration. On a larger and more permanent scale, the work undertaken at the AA's Hooke Park campus also promotes collaboration between designers, engineers, and forestry professionals, leading to a more efficient and sustainable building process using wood. In addition to this, these projects demonstrate an interest in overcoming an extractive approach towards the forest, which is no longer viewed as simply a source of materials but rather as an actor whose capacities and needs must be respected by adapting to them. Viewing the forest as an actor and stakeholder in the wider process of designing and building with wood requires us to consider how to collaborate with it to establish a sustainable building process. The relationship that Hooke Park aims to establish with the forest in which they are situated can be regarded as a model for a deeper collaboration with non-human stakeholders (i.e., the forest and the species within it) and for considering the needs and capacities of the forest when harvesting material.

Projects with a more social focus use cooperation with local stakeholders as one of the main tools to ensure the success of their initiatives. Involving locals in the building process, and collaborating with students, can help embed the projects in the social fabric of the community, favouring a sense of belonging and empowerment. This can be seen in the initiatives of the Insitu project, where the involvement of locals is key to their strategy, which aims to rediscover local timber-building traditions. Working together, students and local communities establish a mutual learning process that is beneficial to both. The former learn about building technologies that support the complex ecological relationship between a community and its living environment; the latter have the opportunity to re-evaluate their knowledge of wood by using it in a modern way, with the help of the university, to preserve their local identity.

Note

1 The research project *Forest Trail* at Umeå School of Architecture is, at the time of going to press, attempting to follow a tree all the way from forest to building. Along the way, the intention is to map all the stakeholders in the production of construction timber. See https://www.umu.se/en/research/projects/forest-trail-global-traces-of-timber-production/.

Bibliography

Council of Europe. 2000. *European Landscape Convention*. Florence: Council of Europe.

Harman, Graham. 2017. *Object-oriented Ontology: A New Theory of Everything*. London: Pelican.

Hutton, Jane Elizabeth. 2018. *Reciprocal Landscapes: Tracing Materials Between New York City and Beyond*. 1st edition. London: Routledge.

Kivinen, Sonja, Jon Moen, Anna Berg, and Åsa Eriksson. 2010. 'Effects of modern forest management on winter grazing resources for reindeer in Sweden.' *Ambio*, 39(4): 269–278. https://doi.org/10.1007/s13280-010-0044-1

Latour, Bruno. 2005. *Reassembling the Social: An Introduction to Actor Network Theory*. Oxford: Oxford University Press.

Magnaghi, Alberto. 2000. *Il progetto locale*. Turin: Bollati Boringhieri.

Magnaghi, Alberto. 2020. *Il principio territoriale*. 1st edition. Turin: Bollati Boringhieri.

Santanicchia, Massimo. 2020. 'Becoming cosmopolitan citizens architects.' *EAAE Annual Conference Proceedings*, 312–335. https://doi.org/10.51588/eaaeacp.71

Chapter 8

A Look Back at the Forest

In Chapter 2 we conceptualised our research as a journey into a forest of study. We entered this forest with certain assumptions about why students, teachers, and architects were choosing to work with wood in schools of architecture, and we approached our eventual case study respondents with certain assertions. Our Scandinavian outlook provided us with a good understanding of how the architecture and construction industry is using timber and engineered timber products. But, through our studies, we discovered a diverse range of alternative attitudes to and applications of wood. Through our forest work, we were challenged, corrected, and enlightened.

We have been careful throughout this book – from the note on terminology through to the chapters and case studies – to distinguish between wood and timber. This has been deliberate. We have found evidence of innovative pedagogical practices that engage students of architecture in a wide range of materials: from largely 'raw' or unprocessed wood to highly sophisticated engineered timber. In writing *Practising Wood*, we have been impressed by the breadth of possibilities available to architectural educators engaging not only with wood but with specific species of wood. Students do not typically begin their architectural education with a confident knowledge of the difference between the wood species that are available to them, and it is self-evident that a hands-on engagement with the material elevates their

DOI: 10.4324/9781003428930-17

Figure 8.1 The forest landscape of inland Västerbotten county, Sweden. Photo: James Benedict Brown.

understanding of a particular type of wood's properties. This runs counter to the tendency to use standardised wood that can be easily quantified and communicated through common grades and measurements of construction timber.

Working hands-on with wood in architectural education is, for many students, a trip into the unknown. In normative architectural education, students do not typically get to build the things they design. Scale models made of paper and cardboard can help to visualise their designs in the studio, and more advanced representations might be possible in the school's workshop. Computers and software can also create millimetre-perfect digital representations. But these are still representations, and it is hard to forget the anecdote that architectural education teaches you how to make drawings of buildings rather than teaching you how to make buildings. To handle the materials used in the construction of buildings and to build something is usually far beyond the capacity, resources, or remit of architectural education. Yet the case studies presented in this book demonstrate the broad range

Practising Wood in Architecture

Figure 8.2 In the quiet hours before the start of the autumn term, the wood workshop at Umeå School of Architecture awaits its first students. Photo: James Benedict Brown.

of possible interactions that students of architecture can have with wood as part of their studies. If the initiatives described in this book have provided some inspiration, what lessons can be drawn from their work?

In each case, we witness a long-term relationship between the initiative and its external stakeholders, whether they are clients, communities, or commercial partners. These initiatives take time and depend on sincere and reciprocal relationships between stakeholders. The material cost of these initiatives is sometimes shared between the participating students, the academic institution, and a commercial partner. Where commercial actors provide materials at a subsidised cost or no cost to the academic institution, a clear understanding of expectations and deliverables has to be communicated.

We also see a broad and critical engagement with the context of education. These initiatives take place in a variety of environmental, cultural, and social contexts. In each case, the initiatives demonstrate an understanding of where they are happening and where the materials being employed are coming from.

For the student, teacher, or architect keen to embark on some kind of hands-on initiative like those described in this book, the provision of physical resources may be a concern. The 'smallest' (in terms of duration and infrastructure) case study in our book is the itinerant Studio in the Woods, which takes place over a long weekend. Although the Studio in the Woods has partnered with well-equipped stakeholders, including landowners and members of the forest industry who have brought equipment for harvesting and milling wood, almost all of its built outcomes have been produced using rechargeable hand tools and manual equipment. Some of the most elegant architectural constructions of previous years have been assembled using modular scaffolding units or simple ratchet straps. With just a few hundred euros worth of equipment, a small workshop can achieve captivating architectural interventions in forest settings.

At the other end of the spectrum, it is hard not to be impressed by the large and high-quality workshop spaces at the well-regarded research universities ETH Zürich and the University of Stuttgart. In these instances, the relationship between teaching and research is fundamental, since the ability to secure large public or private financing for research into architectural materials, construction, or technology can provide long-term investment in the facilities that will also support education. Irrespective of whether the education of architects in the reader's country is funded publicly, by tuition fees, or by a combination of the two, we have seen how well-established design-build initiatives in architectural education exploit the capacity of the university to conduct original scientific research and to embed that research in the education of future architects. But irrespective of the scale and budget available, workshops are nothing without the technicians who staff and maintain them. Susan Orr and Alison Shreeve identify technicians as the more stable (and more frequently available) part of a tripartite relationship with student and tutor, especially in studio-based learning (2018, 79–80). The success of the initiatives studied in this book is directly attributable to the commitment given *by* and the support given *to* technical staff in the academic institution.

Figure 8.3 Pre-fabricating structural components requires large workshop spaces and well-resourced technicians. Photo: Postgraduate Design + Make course, Architectural Association.

For educators who want to work with wood – or work *more* with wood – there are some conclusions to be drawn. To paraphrase Friedrich Nietzsche in *Thus Spoke Zarathustra*: dig deep where you stand (2020, 461–466). Nietzsche invited the reader to engage in critical and introspective exploration: question your beliefs, values, and assumptions, and look at the world around you with fresh eyes. As architects, we interpret Nietzsche's invocation through the additional lens of Sven Lindqvist's manual for popular history *Gräv där du står* (Swedish for a similar phrase: *dig where you stand*). What can you - most likely an architect, architectural teacher, or student – bring to this situation?

All of the case studies in this book demonstrate, to some capacity, a critical engagement not only with the role of the architect, educator, and student but also with the place in which they are working. If you are located in or near woodland or forest, you have a better opportunity than any outsider to engage in a critical understanding of this complex natural environment. Equally, if you are located in a city, you have a better opportunity than any visitor to engage with and understand the complex human environment. It was a relatively benign discovery of one of our doctoral dissertations that architectural educators who engaged in live projects tended to do so in the neighbourhoods and communities where they lived (Brown, 2012). *To dig where you stand* is a natural first step for many architectural educators, students, and practitioners. To work in your local community provides opportunities for deep, meaningful, and long-lasting collaboration.

In several of our case studies, we see evidence of successful trans-national work. In these instances, our invocation to *dig where you stand* might ring hollow. Whilst there is legitimate scepticism of initiatives that exploit students' interest in travel to get them to undertake 'voluntourism' in other countries, we see that design-build initiatives can be socially and economically sustainable provided they secure meaningful and mutual long-term relationships with local stakeholders. Of particular interest is how, at the time of writing this book, the University of Nottingham (UoN) has been working to find a South African university to partner with for its design-build projects. Advanced plans have been made for the first iteration of their projects in which British and South African architecture students work side-by-side on projects in South Africa. If you want to take part in this kind of work, do not presume to be able to do it without the respectful and holistic collaboration of those who know the place in which you are working.

As we undertook this journey through our forest of study, we encountered many new and unfamiliar practices, as well as distinct ecosystems of practice that on occasion bled into one another. Just like a rich, ecologically diverse old-growth forest, we encountered imperceptible changes in the density, mix, and balance of different elements.

We are immensely grateful to the ten educators (from seven countries) who responded to the research informing this book. We have attempted to represent their work accurately and fairly. But perhaps we have misrepresented some aspect of their work or omitted completely another more interesting or relevant case study. With only limited time and resources at our disposal, it is perhaps inevitable that the reader will be the best person to inform us of our oversights and omissions, and we look forward to hearing from you.

If you have been working hands-on with wood with students of architecture, you may recognise some of the themes we have explored regarding the characteristics, capacities, pedagogies, ecologies, and politics of wood. We believe that distinct from its environmental potential, wood is a material with a uniquely transparent ecology, one that can support stakeholders to critically understand the role that materials can play as the products of complex ecological systems and as a catalyst for better perception and understanding of those systems. Wood is more than just a raw material or a solution to the anthropogenic climate emergency; instead, it is the product of many ecologies, from the forest that produced it to the community in which it is eventually used to realise a building. A critical engagement in all of these ecologies is essential to make working with wood a truly sustainable practice.

Bibliography

Brown, James Benedict. 2012. 'A critique of the live project.' Thesis (Ph.D.), Queen's University Belfast.

Lindqvist, Sven. 1978. *Gräv där du står: hur man utforskar ett jobb*. Stockholm: Bonnier.

Nietzsche, Friedrich. 2020. 'Thus spoke Zarathustra.' In *The Routledge Circus Studies Reader*, 461–466. Abingdon: Routledge.

Orr, Susan and Alison Shreeve. 2018. *Art and Design Pedagogy in Higher Education: Knowledge, Values and Ambiguity in the Creative Curriculum*. 1st edition. Abingdon: Routledge.

Further Reading

Anderson, Jane. 2021. 'How can live projects stimulate progress in education, research, and practice? The establishment of live project/design build/community design education as a global and rigorous field of activity and inquiry.' Thesis (Ph.D.), Oxford Brookes University. https://doi.org/10.24384/29e4-h364

Anderson, Jane and Colin Priest. 2014. 'Developing an inclusive definition, typological analysis and online resource for Live Projects.' In *Architecture Live Projects: Pedagogy into Practice*, 9–17.

Anderson, Jane and Priest, Colin. 'Live project network.' Accessed 14 December 2021. https://liveprojectsnetwork.org/about/

Bader, Vera Simone and Andres Lepik (Eds.). 2020. *Experience in Action! DesignBuild in Architecture*. 1st edition. Munich: Edition Detail.

Brown, James Benedict. 2019. *Mediated Space*. 1st edition. Newcastle upon Tyne: RIBA Publishing.

Brown, James Benedict, Harriet Harriss, Ruth Morrow, and James Soane. 2019. *A Gendered Profession: The Question of Representation in Space Making*. 1st edition. Newcastle upon Tyne: RIBA Publishing.

Catsaros, Christophe, Stéphane Berthier, Françoise Fromonot, Yann Rocher, and Yves Weinand. 2020. *Les cahiers de l'Ibois 1*. Lausanne: EPFL Press.

Deplazes, Andrea. 2009. *Constructing Architecture: Materials, Processes, Structures – A Handbook*. Basel: Birkhäuser Verlag.

Gu, Hongmei, Prakash Nepal, Matthew Arvanitis, and Delton Alderman. 2022. 'Carbon impacts of engineered wood products in construction.' In *Engineered Wood Products for Construction*, edited by Meng Gong, 127–138. London: IntechOpen.

Hameury, S. 2005. 'Moisture buffering capacity of heavy timber structures directly exposed to an indoor climate: A numerical study.' *Building and Environment*, 40(10): 1400–1412. https://doi.org/10.1016/j.buildenv.2004.10.017

Hawkins, W., S. Cooper, S. Allen, J. Roynon, and T. Ibell. 2021. 'Embodied carbon assessment using a dynamic climate model: Case-study comparison of a concrete, steel and timber building structure.' *Structures* 33: 90–98. https://doi.org/10.1016/j.istruc.2020.12.013

Hayes, Richard W. 2007. *The Yale Building Project: The First 40 Years*. New Haven, CT: Yale University Press.

Hudert, Markus and Sven Pfeiffer (Eds). 2019. *Rethinking Wood: Future Dimensions of Timber Assembly*. Basel: Birkhäuser Verlag.

Kaminer, Tahl. 2017. *The Efficacy of Architecture: Political Contestation and Agency*. Abingdon: Routledge.

Kaufmann, Hermann and Winfried Nerdinger. 2011. *Building with Timber: Paths into the Future*, edited by Mirjana Grdanjski, Martin Kühfuss, James Roderick O'Donovan, Michael Robinson, and Jennifer Taylor. Munich: Prestel.

Kraus, Chad. 2017. *DesignBuild Education*. London: Routledge.

Krieg, Oliver David, Achim Menges, and Tobias Schwinn. 2017. *Advancing Wood Architecture: A Computational Approach*. New York: Routledge.

Menges, Achim, and Jan Knippers. 2020. *Architecture Research Building: ICD/ITKE 2010/20*. 1st edition. Boston: Birkhauser.

Menges, Achim, Bob Shelll, Ruairi Glynn, and Marilena Skavara. 2017. *Fabricate: Rethinking Design and Construction*. London: UCL Press.

Moe, Kiel. 2017. *Empire, State and Building*. New York: Actar.

Morris, F., S. Allen, and W. Hawkins. 2021. 'On the embodied carbon of structural timber versus steel, and the influence of LCA methodology.' *Building and Environment*, 206. https://doi.org/10.1016/j.buildenv.2021.108285

Mukkavaara, Jani, Marcus Sandberg, Karin Sandberg, Anna Pussette, and Joakim Norén. 2020. 'Sustainability evaluation of timber dwellings in the north of Sweden based on environmental impact and optimization of energy and cost.' *Procedia Manufacturing*, 44: 76–83. https://doi.org/10.1016/j.promfg.2020.02.207

Norman, James and Andrew Thomson. 2020. *Designing Timber Structures: An Introduction*. High Wycombe: BM TRADA Group.

Pak, Burak, and Aurelie De Smet. 2023. *Experiential Learning in Architectural Education: Design-Build and Live Projects*. Abingdon: Routledge.

Prévost, Violaine. 2021. *Les cahiers de l'Ibois 2*. Lausanne: EPFL Press.

Rockhill, Dan and David Sain. 2018. *Studio 804: Design Build – Expanding the Pedagogy of Architectural Education*. Hong Kong: Oscar Riera Ojeda Publishers Limited.

Sara, Rachel. 2006. 'Live project good practice: A guide for the implementation of live projects.' *CEBE Briefing Guide Series, No. 8*. https://advance-he.ac.uk/knowledge-hub/live-project-good-practice-guide-implementation-live-projects-briefing-guide-no-08

Steiger, Ludwig. 2017. *Basics Timber Construction*. 1st edition. Basel: Birkhäuser Verlag.

Storonov, Tolya. 2018. *The Design-Build Studio: Crafting Meaningful Work in Architecture Education*. London: Routledge.

SVT Nyheter Västerbotten. 2023. 'SLU:s nya rön: En månad kortare vinter vid Svartberget i Vindeln.' *SVT Nyheter*. Accessed 13 March 2023. https://www.svt.se/nyheter/lokalt/vasterbotten/en-manad-kortare-vinter-vid-svartberget

Taverna, R., P. Hofer, F. Werner, E. Kaufmann, and E. Thürig. 2007. *The CO_2 Effects of the Swiss Forestry and Timber Industry. Scenarios of future potential for climate-change mitigation*. Bern: Federal Office for the Environment.

Tsing, Anna Lowenhaupt. 2015. *The Mushroom at the End of the World: On the Possibility of Life in Capitalist Ruins*. Princeton, NJ: Princeton University Press.

Weinand, Yves. 2017. *Advanced Timber Structures: Architectural Designs and Digital Dimensioning*. Basel: Birkhäuser Verlag.

Weinand, Yves. 2021. *Design of Integrally-Attached Timber Plate Structures*. 1st edition. London: Routledge.

Winchester, N. and J.M. Reilly. 2020. 'The economic and emissions benefits of engineered wood products in a low-carbon future.' *Energy Economics*, 85. https://doi.org/ARTN 104596.

Zwerger, Klaus and Valerio Olgiati. 2015. *Wood and Wood Joints: Building Traditions of Europe, Japan and China*. 3rd enlarged edition. Basel: Birkhäuser Verlag.

Index

Note: Page entries in *italic* refer to figures and tables, and entries in **bold** refer to case studies.